Do Brilliantly

GCSE Science

Steve Bibby, Phil Hills and Mike Smith

Series Editor: Jayne de Courcy

Published by HarperCollins*Publishers* Limited
77-85 Fulham Palace Road
London W6 8JB

www.**Collins**Education.com
On-line support for schools and colleges

© HarperCollins*Publishers* Ltd 2001

First published 2001

ISBN 0 00 710491 X

Steve Bibby, Phil Hills and Mike Smith assert the moral right to be identified as the authors of this work.

British Library Cataloguing in Publication Data
A catalogue record for this book is available from the British Library

Edited by Eva Fairnell
Production by Kathryn Botterill
Cover design by Susi Martin-Taylor
Book design by Gecko Limited
Printed and bound by Scotprint Ltd, Haddington

Acknowledgements
The Authors and Publishers are grateful to the following for permission to reproduce copyright material:
AQA 9; CCEA 14, 28, 29; NEAB, a division of AQA, 12, 17; OCR 13, 18, 19, 22, 23, 24, 27, 32, 33, 34, 37, 38, 62, 63, 64, 74, 77, 79, 84, 92, 93, 99, 107, 108; WJEC 54, 83, 94, 102.
Answers to questions taken from all past examination papers are entirely the responsibility of the authors and have neither been provided nor approved by the examination board.

Illustrations
Cartoon Artwork – Dave Mostyn and Roger Penwill

Photographs
Allsport 86; Andrew Lambert 36; Science Photo Library 11, 16, 21, 26, 41, 46, 51, 56, 61, 66, 71, 76, 91, 96, 101, 106; Sony 81; Tony Stone Images 31.

You might also like to visit:
www.**fire**and**water**.com
The book lover's website

Contents

How this book will help you

by Steve Bibby, Phil Hills and Mike Smith

Exam practice – how to answer questions better

This book will help you improve your performance in your GCSE Science exams (Double or Single Award). In Science exams, examiners are looking to reward good answers and will not deduct marks for everything you write that is incorrect. However, to gain as many marks as you can, you need to make sure that what you write:

- **is scientifically correct**
- **is in sufficient detail**
- **answers the question.**

Every year many exam candidates don't use the information they've learnt as effectively as they could, and so they don't get the grade that they're probably capable of achieving. **You will get a high GCSE Science grade through a combination of good knowledge, good understanding and good examination technique. It's the last of these that this book will particularly help you to improve.**

Each chapter in this book is broken down into four separate sections:

❶ Exam Question and Answer and 'How to score full marks'

The exam questions at the start of each chapter have been chosen to show typical questions for that topic area. We've also chosen them to illustrate some of the **different styles of questions** you may come across. The answers are 'model' answers, in other words, answers that would gain full marks.

In the 'How to score full marks' section we have included comments that:
- explain what questions are asking
- highlight common errors that would not gain marks
- show other ways of gaining full marks.

When you meet these sorts of questions in your exams you will know exactly how to answer them and gain full marks.

❷ 'Don't make these mistakes'

The 'Don't make these mistakes' section in each chapter highlights the common mistakes that examiners such as ourselves see every year in students' exam papers. These include common scientific misconceptions and mistakes students make when answering questions.

When you're into your last minute revision, you can quickly read through all these sections to make doubly sure that you avoid these mistakes in your exam.

❹ Questions to try, Answers and Examiner's Hints

Each chapter ends with exam questions for you to try answering. Don't cheat. Sit down and try to answer the question as if you were in an exam. Try to remember all that you've read earlier in the chapter and put it into practice. Look at the **Examiner's Hints** before you answer.

Check your answer through and then look at the answers given at the back of the book. The answers are based on the mark schemes that examiners would use to mark the questions. They show all the answers that would gain marks. **The sections after each answer highlight what is needed to gain full marks as well as common errors to avoid.** Compare your answer with the answers given and decide whether you would have gained full marks and, if not, what you need to improve on.

❸ 'Key points to remember'

A book of this size can't cover all the scientific information that you will need to do well in your Science exam. **The 'Key points to remember' section lists the most important points that you need to cover when revising a particular topic.** It does not include all the detail that you will need to know – you'll need to use your notes and textbook as well for this. Remember that **the detail** of your answers is often what is needed to achieve the very highest marks and hence the top grades of A or A*.

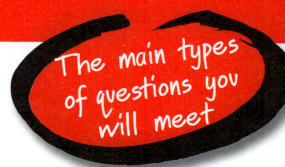
The main types of questions you will meet

This book contains questions from different Exam Groups and different syllabuses as well as questions especially written to illustrate important points. Each Exam Group syllabus has particular styles of questions and this is a good reason for getting hold of as many past papers for your Exam Group syllabus as you can.

This book is divided into 20 topic-based chapters covering the main topics in GCSE Science. **Although each Exam Group syllabus is slightly different, the main topics will be the same whichever syllabus you are studying.**

The skills being tested

Different exam questions will be testing different things:
- your **knowledge** and **understanding** of scientific facts, ideas and processes
- your ability to **use** and **apply** your scientific knowledge in different situations
- your ability to **communicate** scientific ideas, for example using scientific language, chemical symbols, mathematical formulae and graphs
- your ability to **evaluate** scientific information and use it to make **judgements**, for example you might be asked to use some experimental results to predict the outcome of further experiments.

There are also different types of exam questions, such as:

Multiple-choice questions

These are questions where you are given the possible answers and have to choose the correct one(s). The questions may be set out in different ways, for example, you may be asked to label a diagram and are given a list of words to choose from.

In some cases you may have to use the same answer for more than one question.

To give yourself the best chance of gaining marks, eliminate the answers you know are not correct and then choose from those that are left.

Extended answer questions

These are questions where you have to write at some length. They will have more than one mark allocated, perhaps up to four or five. You will also have much more space to answer, depending on how many marks are allocated, perhaps up to ten lines, or even more. In this type of question you might have to, for example, describe a process in detail or explain some experimental results.

To gain full marks, make sure you answer in sufficient detail, making at least as many points as there are marks.

Short answer questions

These are questions where there is usually one mark allocated and one or two lines space to answer. **Often you can correctly answer these questions with only a few words**; in some cases a single word may be enough.

Make sure your answers are clear and to the point.

Questions across different topic areas

The chapters in this book are topic-based and so are the questions in them. However, many exam questions will not be about a single topic, but rather will be testing your knowledge of **different topic areas within the same question**. In some exams you may even have different parts of the same question testing ideas from Biology, Chemistry and Physics! Work through these questions particularly carefully; make sure you know which topic/subject area each part of the question is testing and then apply the techniques this book teaches you about how to gain full marks.

Exam tips

- **Read each question carefully**; this includes looking in detail at any **diagrams**, **graphs** or **tables**. Remember that any information you are given is there to help you to answer the question. Underline or circle the **key words** in the question and **make sure you answer the question that is being asked** rather than the one you wish had been asked!

- Make sure that you understand the meaning of the '**command words**' in the questions, for example:
  - '**Describe**' is used when you have to give the main feature(s) of, for example, a process or structure;
  - '**Explain**' is used when you have to give reasons, e.g. for some experimental results;
  - '**Suggest**' is used when there may be more than one possible answer, or when you will not have learnt the answer but have to use the knowledge you do have to come up with a sensible one;
  - '**Calculate**' means that you have to work out an answer in figures.

- Look at the **number of marks** allocated to each question and also the **space provided** to guide you as to the length of your answer. You need to make sure you include at least as many points in your answer as there are marks, and preferably more. If you really do need more space to answer than provided, then use the nearest available space, e.g. at the bottom of the page, making sure you write down which question you are answering. **Beware of continually writing too much because it probably means you are not really answering the questions**.

- Don't spend so long on some questions that you don't have time to finish the paper. You should spend approximately **one minute per mark**. If you are really stuck on a question, leave it, finish the rest of the paper and come back to it at the end. Even if you eventually have to guess at an answer, you stand a better chance of gaining some marks than if you leave it blank.

- In short answer questions, or multiple-choice type questions, **don't write more than you are asked for**. In some exams, examiners apply the rule that they only mark the first part of the answer written if there is too much. This means that the later part of the answer will not be looked at. In other exams you would not gain any marks, even if the first part of your answer is correct, if you've written down something incorrect in the later part of your answer. This just shows that you haven't really understood the question or are guessing.

- **In calculations always show your working out**. Even if your final answer is incorrect you may still gain some marks if part of your attempt is correct. If you just write down the final answer and it is incorrect, you will get no marks at all. Also in calculations you should write down your answers to as many **significant figures** as are used in the question. You may also lose marks if you don't use the correct **units**.

- In some questions, particularly short answer questions, answers of only one or two words may be sufficient, but in longer questions you should aim to use **good English** and **scientific** language to make your answer as clear as possible.

- If it helps you to answer clearly, don't be afraid to also use **diagrams** or **flow charts** in your answers.

- When you've finished your exam, **check through** to make sure you've answered all the questions. Cover over your answers and read through the questions again and check your answers are as good as you can make them.

The Periodic Table

Key:

atomic mass
symbol
name
atomic no.

non metal	metal

Group	1	2											3	4	5	6	7	0 or 8
Period 1	1 **H** hydrogen 1																	4 **He** helium 2
Period 2	7 **Li** lithium 3	9 **Be** beryllium 4											11 **B** boron 5	12 **C** carbon 6	14 **N** nitrogen 7	16 **O** oxygen 8	19 **F** fluorine 9	20 **Ne** neon 10
Period 3	23 **Na** sodium 11	24 **Mg** magnesium 12											27 **Al** aluminium 13	28 **Si** silicon 14	31 **P** phosphorus 15	32 **S** sulphur 16	35.5 **Cl** chlorine 17	40 **Ar** argon 18
Period 4	39 **K** potassium 19	40 **Ca** calcium 20	45 **Sc** scandium 21	48 **Ti** titanium 22	51 **V** vanadium 23	52 **Cr** chromium 24	55 **Mn** manganese 25	56 **Fe** iron 26	59 **Co** cobalt 27	59 **Ni** nickel 28	64 **Cu** copper 29	65 **Zn** zinc 30	70 **Ga** gallium 31	73 **Ge** germanium 32	75 **As** arsenic 33	79 **Se** selenium 34	80 **Br** bromine 35	84 **Kr** krypton 36
Period 5	85.5 **Rb** rubidium 37	88 **Sr** strontium 38	89 **Y** yttrium 39	91 **Zr** zirconium 40	93 **Nb** niobium 41	96 **Mo** molybdenum 42	98 **Tc** technetium 43	101 **Ru** ruthenium 44	103 **Rh** rhodium 45	106 **Pd** palladium 46	108 **Ag** silver 47	112 **Cd** cadmium 48	115 **In** indium 49	119 **Sn** tin 50	122 **Sb** antimony 51	128 **Te** tellurium 52	127 **I** iodine 53	131 **Xe** xenon 54
Period 6	133 **Cs** caesium 55	137 **Ba** barium 56	139 **La** lanthanum 57	178.5 **Hf** hafnium 72	181 **Ta** tantalum 73	184 **W** tungsten 74	186 **Re** rhenium 75	190 **Os** osmium 76	192 **Ir** iridium 77	195 **Pt** platinum 78	197 **Au** gold 79	201 **Hg** mercury 80	204 **Tl** thallium 81	207 **Pb** lead 82	209 **Bi** bismuth 83	210 **Po** polonium 84	210 **At** astatine 85	222 **Rn** radon 86
Period 7	223 **Fr** francium 87	226 **Ra** radium 88	227 **Ac** actinium 89	261 **Db** dubnium 104	262 **Jl** joliotium 105	**Rf** rutherfordium 106	**Bh** bohrium 107	**Hn** hahnium 108	**Mt** meitnerium 109									

Lanthanides:

140 **Ce** cerium 58	141 **Pr** praseodymium 59	144 **Nd** neodymium 60	147 **Pm** promethium 61	150 **Sm** samarium 62	152 **Eu** europium 63	157 **Gd** gadolinium 64	159 **Tb** terbium 65	162.5 **Dy** dysprosium 66	165 **Ho** holmium 67	167 **Er** erbium 68	169 **Tm** thulium 69	173 **Yb** ytterbium 70	175 **Lu** lutetium 71

Actinides:

232 **Th** thorium 90	231 **Pa** protactinium 91	238 **U** uranium 92	237 **Np** neptunium 93	242 **Pu** plutonium 94	243 **Am** americium 95	247 **Cm** curium 96	247 **Bk** berkelium 97	251 **Cf** californium 98	254 **Es** einsteinium 99	253 **Fm** fermium 100	256 **Md** mendelevium 101	254 **No** nobelium 102	257 **Lr** lawrencium 103

Exam Question and Answer

1) a) The diagram shows four ways in which molecules may move into and out of a cell. The dots show the concentration of molecules.

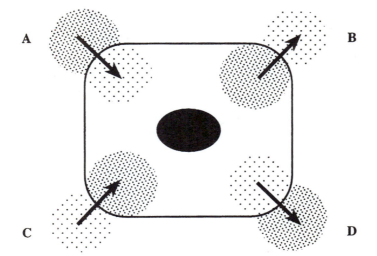

The cell is respiring aerobically.

Which arrow, **A**, **B**, **C** or **D**, represents:

i) movement of oxygen molecules

A (1 mark)   [1 mark]

ii) movement of carbon dioxide molecules?

B (1 mark)   [1 mark]

b) Name the process by which these gases move into and out of the cell.

Diffusion (1 mark)   [1 mark]

c) Which arrow, **A**, **B**, **C** or **D**, represents the active uptake of sugar molecules by the cell?

C (1 mark)

Explain the reason for your answer.

There is a high concentration inside the cell and a low concentration outside the cell. Usually molecules move from a high concentration to a low concentration but here they are going in the opposite direction so it must be active uptake (1 mark).   [2 marks]

How to score full marks

a) Because the cell is respiring aerobically it will be using up oxygen. This means the oxygen concentration will be **lower inside** the cell than outside. This will therefore cause oxygen to **enter** the cell (by diffusion). Choosing **A** for **(i)** shows that you understand these ideas.

Carbon dioxide will be produced by the cell during aerobic respiration, increasing the concentration inside the cell. As the concentration inside the cell is **higher** than outside, the carbon dioxide will diffuse **out** of the cell. Therefore the answer to **(ii)** must be **B**.

Even if you are stuck on a question like this in an exam and are forced to guess, you should at least be able to narrow down the choices.

b) You should know the different ways that substances move in and out of cells. The answer cannot be osmosis because that only applies to water. Nor is it likely to be active transport because the gases are moving from high to low concentrations.

The answer to (b) is also unlikely to be active transport because (c) mentions active transport in the question and an answer would not usually be given away like that!

c) Active uptake (active transport) can either move substances from a high to a low concentration (helping to speed up the movement that will already be happening by diffusion) or in the opposite direction to diffusion. In exam questions, however, **when active transport is mentioned it is usually an example where molecules are moving from a low to a high concentration and so it can't be anything else.**

Don't make these mistakes...

Don't confuse a cell membrane with a cell wall; they are not the same thing. Remember that all cells, plant and animal, have a cell membrane but that plant cells have a cell wall as well.

Don't describe plant cells as being rectangular in shape. It is true that plant cells are often shown like this in textbooks, but this is just to make the diagram simpler. (Also don't forget that in reality they are three-dimensional.)

Don't confuse osmosis and diffusion. Diffusion is the movement of particles from a high to a low concentration. Osmosis is the movement of water, across a partially permeable membrane, from a more dilute solution (i.e. a high concentration of water) to a more concentrated solution (i.e. a low concentration of water). It is true that osmosis is a special type of diffusion but osmosis only applies to water.

Don't confuse respiration with breathing. Respiration is the way energy is released from food. Breathing is the way that air is taken in and out of the lungs.

Don't confuse active transport and diffusion. Active transport needs energy from the cell to move substances across the cell membrane. Diffusion does not use energy. Active transport can move substances from a high to a low concentration (just like diffusion) but can also move them from a low to a high concentration (i.e. in the opposite direction to diffusion).

Don't confuse aerobic and anaerobic respiration. Aerobic respiration uses oxygen to release energy from food. Anaerobic respiration takes place when there is a lack of oxygen. There are also differences in the waste products and the amount of energy released.

Cells may be specialised to do different jobs but they have many features in common:
- a nucleus that contains genetic information
- a cell membrane that controls substances entering and leaving the cell
- cytoplasm in which many chemical processes take place
- mitochondria where respiration occurs.

In addition plant cells have some features not found in animal cells:
- a cell wall that provides support
- chloroplasts that contain chlorophyll to absorb sunlight for photosynthesis
- a large vacuole (containing sap) which helps to support the cell.

Diffusion is the movement of gas or liquid particles from an area of high concentration to an area of lower concentration. One place that diffusion happens is in the lungs where oxygen diffuses from the alveoli (air sacs) in to the blood. Carbon dioxide diffuses from the blood in to the alveoli.

Osmosis is a special type of diffusion in which water moves from a dilute solution to a more concentrated one through a partially permeable membrane. One example of osmosis is water entering plant root hairs from the soil.

Active transport is another way in which substances can move across a cell membrane. It involves carrier molecules in the membrane itself moving the substances from one side to the other. The cells need to use energy to do this. Active transport can move substances from a low to a high concentration, which is in the opposite direction to diffusion. An example of this is the absorption of minerals, e.g. nitrates, from the soil by plant root hairs.

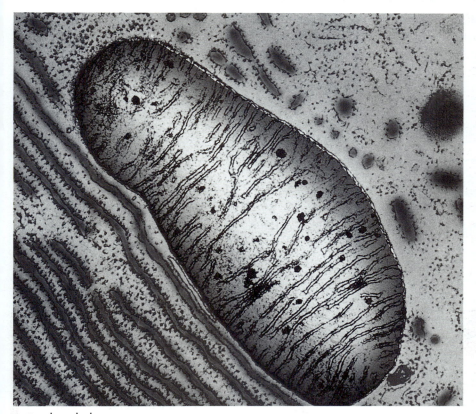

Animal and plant cells have some cell organelles in common. All cells have mitochondria.

All living things need energy, which they get from food. The way in which energy is released is called respiration. Respiration happens inside the cells of each living thing.

Aerobic respiration uses oxygen to release energy:

glucose + oxygen → carbon dioxide + water + energy

Anaerobic respiration releases energy from food when there is a lack of oxygen. It is not as efficient as aerobic respiration and releases less energy from each glucose molecule.

An example of anaerobic respiration is in muscle cells during exercise when they may not be able to get oxygen quickly enough to provide all they energy they need:

glucose → lactic acid + energy

Lactic acid causes muscle ache and extra oxygen will be needed after the exercise to break it down. This is called 'repaying the oxygen debt'.

Questions to try

Examiner's Hints

● The sugar has changed to a syrup, which means it has gained water. Think about where the water has come from and how it moved.

● There are 4 marks available so your answer should include at least 4 ideas. You may not need to fill up all the available space when answering but you will probably need to use most of it to gain full marks.

A cook prepares a fresh fruit salad by cutting up a variety of fruits and placing them in a bowl with layers of sugar in between. After two hours the fruit is surrounded by syrup (concentrated sugar solution).

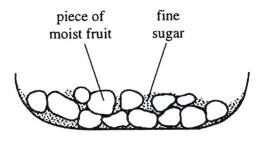

Freshly-prepared fruit salad   Two hours later

Explain, as fully as you can, why syrup (concentrated sugar solution) was produced after two hours.

_____

_____

_____

_____

_____

_____

_____

_____ [4 marks]

Examiner's Hints

(a) This question is not just about respiration. You will have to use ideas about breathing, the blood system and homeostasis.

(b) Think about what is happening in (a) to help you answer (b)(i). (ii) and (iii) are straight forward if you have learnt the facts about anaerobic respiration.

(a) Our bodies use oxygen to release energy.

This is called aerobic respiration. When you exercise you breathe faster to take in more oxygen. What happens in your body to control the increase in breathing rate? Explain as fully as you can.

_____

_____

_____

_____ [4 marks]

(b) During **hard** exercise the muscles cannot get enough oxygen to use to release all the energy they need.

  (i) Why can't the muscles get enough oxygen during hard exercise?

  _____ [1 mark]

  (ii) The muscles also use anaerobic respiration to release the extra energy they need. Complete the word equation for anaerobic respiration in the muscles.

  _____ → _____ + energy   [2 marks]

  (iii) Anaerobic respiration does not use oxygen. Write down **one** other way that anaerobic respiration is different from aerobic respiration.

  _____ [1 mark]

Answers are given on p.109

Exam Question and Answer

1) The diagram below shows the structure of a human heart.
The four chambers are labelled **A**, **B**, **C** and **D**.

a) Which letter (**A–D**) represents:

    i)   the left ventricle   *D* **(1 mark)**   [1 mark]

    ii) the right atrium?   *A* **(1 mark)**   [1 mark]

b) **i)** Blood is made up of **four** different components. Two of them are
      white blood cells and **platelets**. Name the other **two**.

      1 *Red cells* **(1 mark)**

      2 *Plasma* **(1 mark)**   [2 marks]

  ii) What is the main function of:
      1 White blood cells?

      *To protect the body from infection by destroying any invading
      microbes such as harmful bacteria or viruses* **(1 mark)**.   [1 mark]

      2 Platelets?

      To make the blood clot to stop any bleeding if you're cut **(1 mark)**. [1 mark]

  iii) Heart attacks can be caused by the blood vessels leading to
      the heart becoming blocked. What **two** precautions can we
      take to help prevent this?

      1 *Cut down on fatty foods* **(1 mark)**.

      2 *Don't smoke* **(1 mark)**.   [2 marks]

How to score full marks

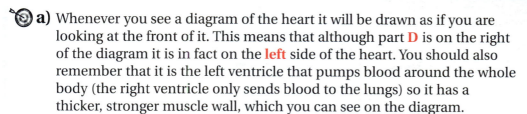

a) Whenever you see a diagram of the heart it will be drawn as if you are looking at the front of it. This means that although part **D** is on the right of the diagram it is in fact on the **left** side of the heart. You should also remember that it is the left ventricle that pumps blood around the whole body (the right ventricle only sends blood to the lungs) so it has a thicker, stronger muscle wall, which you can see on the diagram.

The right atrium is where blood collects before entering the right ventricle.

In questions like this, where there is only one possible answer for each part make sure you only put down *one* answer. If you put down more it shows that you are not sure.

b) **i)** A common error would be to forget the plasma. This is not just simply the liquid that the blood cells are carried in, it also carries many important dissolved substances.

You have been asked for two answers, so in this question make sure you only give two. If you give more you could lose marks even if you include the correct answers. A common rule of marking for examiners would be, in this case, to only mark the first two answers you give.

ii) **Make sure you give enough detail to be sure of getting the marks.** Shorter answers than the examples given might be sufficient but don't take chances in an exam.

iii) There are other precautions (worth 1 mark each) that can help prevent heart disease, such as cutting down on alcohol, reducing stress levels, taking plenty of exercise.

Don't make these mistakes...

Don't confuse how substances are carried in the blood. Oxygen is carried by red blood cells but almost everything else is carried dissolved in the plasma.

Don't confuse breathing (ventilation), gaseous exchange and respiration (see Chapter 1).

Don't confuse egestion and excretion (see Chapter 3). Excretion is removing waste substances that have been produced by the body (e.g. carbon dioxide or urea). Getting rid of faeces is described as egestion. The waste in faeces has not been produced by the body, but is what is left of the food after useful substances have been absorbed into the blood.

Don't say that when we breathe in, the air entering is what makes the lungs expand. Also do not describe the air as being sucked in. Both ideas are wrong.

If you are describing chemical digestion, do not forget to mention that it is the food molecules themselves that are being broken down into smaller ones.

Don't say that enzymes have been killed by high temperatures (or extremes of pH). They are not alive so can't be killed. Instead say they have been denatured.

Don't say that stomach acid digests food. It does provide the right pH for the stomach enzymes to work (and also kills harmful microbes) but this is not the same thing.

Don't confuse digestion with eating. Digestion is the breakdown of food into a soluble form so it can be absorbed into the blood.

Key points to remember

The main function of the blood system is to transport substances around the body.

Blood contains:
- red blood cells to carry oxygen
- white blood cells to fight disease
- platelets to help blood clot
- plasma (liquid) to carry substances such as food, waste, hormones.

The blood (circulatory) system is made up of:
- the heart to pump blood
- arteries, which have thick walls to deal with the high pressure of the blood as they carry it away from the heart
- veins, which have thinner walls (and valves) and carry blood at a lower pressure back to the heart
- capillaries, which have permeable walls to allow substances to enter and leave the blood to exchange materials with body tissues.

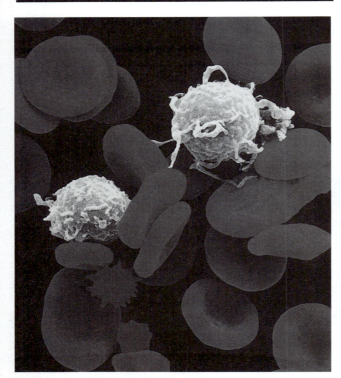

Red blood cells have a distinctive biconcave shape. White blood cells have a more varied shape.

The main function of the breathing system is to provide oxygen for respiration and remove carbon dioxide.

It is made up of the mouth and nose, trachea (windpipe), bronchi, bronchioles and alveoli (air sacs).

Breathing or ventilation is the movement of air in and out of the lungs using the rib cage and diaphragm. These move to change the volume inside the lungs, which changes the air pressure inside the lungs causing air to enter or leave.

Gaseous exchange is the exchange of oxygen and carbon dioxide between the air in the alveoli and the blood.

Alveoli are adapted for efficient gas exchange by:
- providing a large surface area
- being moist, thin and permeable
- having a good blood supply.

The main function of the digestive system is to break down (digest) food so it can enter the blood and be carried around the body.

Physical (mechanical) digestion is breaking pieces of food into smaller ones e.g. by chewing.

Chemical digestion is breaking large insoluble food molecules into smaller, more soluble ones.

Chemical digestion is carried out by enzymes, which:
- are specific (each enzyme digests one type of food)
- work best at their own optimum temperature and pH
- can be denatured (irreversibly damaged) by high temperatures or too high or low pHs.

The main parts of the digestive system are the:
- mouth, to chew food and start the digestion of carbohydrates
- oesophagus, to carry food to the stomach by peristalsis (muscle movements)
- stomach, to store food and start the digestion of protein
- gall bladder, to store and release bile, which emulsifies fats
- pancreas, which secretes many different enzymes
- small intestine (duodenum and ileum), where digestion is completed and digested food is absorbed into the blood through the villi
- large intestine (colon), where water is absorbed into the blood
- rectum, which compacts and stores waste (faeces)
- anus, where faeces is egested.

Examiner's Hints

(a) Think about how the food is digested as it passes through the digestive system, and what happens after it has been digested. Note that the question is specifically about protein.

(b) Explain the results shown in **each** line of the table. Use information from the table as well as your own knowledge of enzymes.

(a) A food contains protein. Describe, in as much detail as you can, what happens to this protein after the food is swallowed.

_____

_____

_____

_____

_____

_____

_____

_____ [4 marks]

(b) The table shows the activity of lipase on fat in three different conditions.

Condition	Units of lipase activity per minute
Lipase + acid solution	3.3
Lipase + weak alkaline solution	15.3
Lipase + bile	14.5

Explain, as fully as you can, the results shown in the table.

_____

_____

_____

_____

_____ [3 marks]

Look at the diagram. It shows part of the breathing system.

rib cage

lung

diaphragm

(a) Look at the list.

> alveolus
> bronchiole
> bronchus
> larynx

Complete the labels on the diagram. Choose your answers from the list.   [2 marks]

(b) When we breathe in, the rib cage and diaphragm move.
How do the rib cage and diaphragm make us inhale air? Explain as fully as you can.

_____

_____

_____

_____

_____ [4 marks]

Answers are given on p.110

Exam Question and Answer

1) Petra caught chickenpox when she was five.

   She was ill for about six days. Then she got better.

   Her friend Roshan caught chickenpox several weeks later.

   They played together a lot when Roshan was ill.

   Petra did not catch chickenpox again.

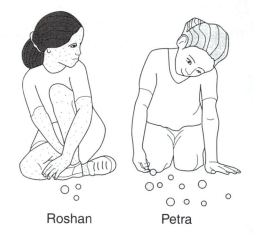

Roshan   Petra

a) Explain why Petra did not catch chickenpox a second time.

   When she caught chickenpox, Petra's body produced antibodies **(1 mark)** which stuck to the antigens **(1 mark)** on the chickenpox germs to destroy them. Her body remembered how to make the antibodies **(1 mark)** so if she caught the chickenpox germ again it would be destroyed before she developed the illness.

   [3 marks]

Petra has now grown up.

She decides to go to Africa for a holiday.

She goes to her doctor for an injection to protect her against typhoid.

Typhoid is a serious disease. It is common in some parts of Africa.

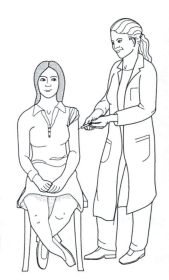

b) The doctor says that the injection contains typhoid antigens, but Petra should not worry. Explain why the injection is not dangerous.

   The typhoid antigens are on some dead typhoid germs **(1 mark)**. Because the germs are dead they cannot multiply and give her the disease **(1 mark)**. [2 marks]

c) Petra decides to extend her holiday and go to the Far East.
   The doctor tells her that there are other diseases that she could catch there.
   She will need some different injections.

   Explain why the typhoid injection will not protect Petra against these other diseases.

   The typhoid injection lets her body make typhoid antibodies but these will only attack the antigens on typhoid germs **(1 mark)**. Other diseases are caused by different germs which will have different antigens **(1 mark)** so different antibodies will be needed to fight these **(1 mark)**. [3 marks]

How to score full marks

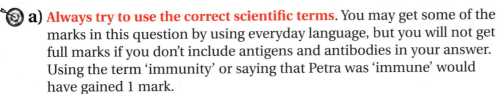**a)** **Always try to use the correct scientific terms**. You may get some of the marks in this question by using everyday language, but you will not get full marks if you don't include antigens and antibodies in your answer. Using the term 'immunity' or saying that Petra was 'immune' would have gained 1 mark.

b) You are not expected to know in detail about typhoid specifically, but **you should be able to explain that the antigens are harmless because Petra has been given a dead, inactive or weakened form** of the typhoid microbe. The second mark is for explaining that the typhoid microbes will not multiply and spread and so cause the disease. You will only develop an illness if the harmful microbe reproduces and spreads.

Make sure you make at least as many points in your answer as there are marks for the question.

c) The question is still about antigens and antibodies **but marks will not be given for the same points as before**, so don't be tempted to repeat information given earlier. Again, you have to make clear points in an extended answer to gain full marks.

Don't make these mistakes ...

Don't say that the job of the myelin sheath of a neurone is to protect the cell. It acts as insulation and also helps to speed up the transmission of impulses.

Don't say that a reflex is controlled by the brain. Part of the central nervous system will be involved but in many reflexes it is the spinal cord because it is nearer. Signals may also be sent to the brain so that it is aware of what is happening, but this is not the same thing as controlling the reflex.

Don't say that erect hairs on the skin keep us warm by trapping a layer of warm air. The point is that they trap a layer of air and air is an insulator.

Don't confuse glucose, glycogen and glucagon. Insulin converts excess glucose to glycogen. Glucagon is a hormone that does the opposite.

Don't confuse the ureters with the urethra. Ureters carry urine from the kidneys to the bladder. The urethra carries urine from the bladder to the outside.

Don't say that we sweat to lose excess water. We sweat to stop us overheating. If we lose too much water this way then the kidneys will produce less urine to compensate.

If you are describing how mucus and cilia (tiny hairs) in the airways of the lungs trap and remove air-borne pathogens and dirt, don't say that the cilia trap them. It is the mucus that does this. The cilia move them up to the top of the oesophagus.

Don't say that sweat works because it is cold. Sweat will be the same temperature as your skin. It cools us down because it takes heat from the skin to allow it to evaporate.

Don't say that during vasodilation blood vessels in the skin 'move closer to the surface'. What happens is that more blood flows in the blood vessels closer to the surface. The blood is warm so heat is lost from the skin more easily.

Key points to remember

The nervous system is made up of:
- the central nervous system (CNS), i.e. the brain and spinal cord which process information
- the peripheral nerves, which carry information around the body.

Nerves are made of cells called neurones:
- sensory neurones carry information from receptors (e.g. sense organs like the eye) to the CNS
- motor neurones carry information from the CNS to effectors (muscles or glands).

Information passes along nerves as electrical impulses. Neurones are connected at synapses where chemical transmitter substances carry the signals between neurones.

Different parts of neurones are:
- the cell body, containing the nucleus
- the myelin sheath, which insulates the neurone and speeds up the impulses
- dendrites, which pick up impulses from other neurones.

Reflexes are automatic responses to stimuli.

The body defends itself against the entrance of pathogens (e.g. harmful bacteria or viruses) in several ways:
- skin is a natural barrier
- mucus lining the airways traps pathogens which the cilia remove
- stomach acid
- blood clots prevent infection.

The white blood cells of the immune system attack pathogens inside the body:
- phagocytes engulf and digest them
- lymphocytes produce antibodies that destroy pathogens by binding to antigens on their surface.

Each type of pathogen carries different antigens and a specific antibody is needed for each. We become immune to diseases when the body has 'learnt' to make the right type of antibody.

The endocrine (hormone) system is made up of endocrine glands that secrete hormones into the blood which carries them to their target organs.
- Adrenaline is produced by the adrenal glands. It prepares the body for 'fight or flight' responses.
- Insulin is produced by the pancreas. It travels to the liver and converts excess glucose in the blood into glycogen, which is stored in the liver until needed.
- Growth hormone is produced in the pituitary gland. It encourages mental and physical development.
- Testosterone is the male sex hormone. It is produced in the testes. It controls sperm production and the secondary sexual characteristics, e.g. hair growth on face and body, and voice breaking.
- Oestrogen and progesterone are the female sex hormones. They are produced by the ovaries. They control the menstrual cycle and secondary sexual characteristics, e.g. breast development and hip widening.

Homeostasis means maintaining the body's internal conditions.

Temperature is controlled by the hypothalamus in the brain, which monitors blood temperature and controls the body's responses to keep body temperature at 37°C:
- vasodilation or vasoconstriction controls how much warm blood flows near the skin surface
- sweating loses heat through evaporation
- hairs stand on end to trap a layer of insulating air
- shivering generates heat.

Water is gained in the body from food, drink and respiration. Water is lost in urine, sweat, breathing and faeces. Water balance is maintained by the hypothalamus, which controls how much water the kidneys remove from the blood to be lost in urine. Urine also contains the waste substance urea. Removing waste produced in the body is called excretion. The lungs removing carbon dioxide is another example of excretion.

Homeostasis involves negative feedback mechanisms.

Substances such as alcohol, tobacco and solvents can harm the body. Increased use can build up the body's tolerance so larger amounts are taken. Some drugs are addictive.

Polar bears have adapted to maintain their internal body conditions in a very severe habitat

Questions to try

The diagram shows the junction of two neurones (nerve cells).

sensory neurone

motor neurone

(a) Draw an arrow, on the diagram, to show the direction of the nerve impulses.   [1 mark]

(b) Write down the name given to the junction of two neurones (nerve cells).

_____ [1 mark]

(c) Describe how the nerve impulses travel across the junction.

_____

_____ [2 marks]

(d) Suggest a substance that will affect the way an impulse travels across a junction.

_____ [1 mark]

(e) The diagram shows the pathway of cells for a simple nervous response.
   (i)   Finish the diagram by writing the correct words in the boxes.

   | stimulus | → | | → | neurones | → | | → | response |

   [2 marks]

   (ii) Write down **one** reason why reflex actions are important to the body.

_____ [1 mark]

Questions to try

Examiner's Hints

(a) You will need to use the graph as well as your knowledge of the female sex hormones.
(b) Think about what else happens in the ovaries apart from hormone production.
(c) There are several possible answers. You can pick up clues to some answers from the graph.
(d) A detailed answer will be needed to gain full marks. Include the name of the hormone, where it works and what exactly it does.

The graph shows the amount of two sex hormones in the blood of a woman.

The hormones are made in the ovaries.

The graph also shows the thickness of the uterus lining.

The growth of the uterus lining is affected by the hormones.

amount of hormones in blood

hormone X

hormone Y

thickness of uterus wall

menstrual bleeding

0   7   14   21   28

days

(a) Look at the graph.

(i)   Write down the name of hormone **X**. _____ [1 mark]

(ii)   Write down the name of hormone **Y**. _____ [1 mark]

(b) Female sex hormones can have an effect on the ovaries. Describe **one** effect which a female sex hormone has on the ovaries.

_____ [1 mark]

(c) Some women regularly take tablets containing female sex hormones. Why do some women take these tablets?

_____ [1 mark]

(d) Other hormones control different processes in our bodies.
The pancreas produces a hormone which helps to control the amount of glucose in the blood.
Write about how this hormone controls the amount of glucose in the blood.

_____

_____

_____ [4 marks]

Answers are given on p.111

Exam Question and Answer

1) Keith is investigating transpiration in a plant. He waters a plant in a plant pot. He wraps the pot in thin plastic film and puts it onto a mass balance. The graph shows the mass of the potted plant during the day.

pot wrapped in plastic

a) Suggest why Keith wrapped the pot in plastic film.

To stop water evaporating from the soil. **(1 mark)** [1 mark]

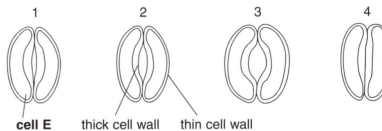

Diagrams **1**, **2**, **3** and **4** show stomata from the leaves of Keith's plant.

The diagrams show the stomata at the points marked **A**, **B**, **C** and **D** on the graph but they are **not** shown in the correct order.

b) Look at the diagrams and the graph. Write down the name of **cell E**.

Guard cell **(1 mark)** [1 mark]

cell E   thick cell wall   thin cell wall

c) Suggest which diagram shows a stoma at **point A** on the graph. Choose from **1**, **2**, **3** or **4**.

3 **(1 mark)**

Explain your answer.

At point A the plant is losing water most quickly. This would happen when the stoma was at its widest, which is number 3 **(1 mark)**. [2 marks]

d) i) Photosynthesis can happen in **cell E**. Complete the symbol equation for photosynthesis.

$$6CO_2 + 6H_2O \xrightarrow[\text{light}]{\text{chlorophyll}} C_6H_{12}O_6 \text{ (1 mark)} + 6O_2$$

[1 mark]

ii) Photosynthesis in **cell E** may cause water to move into the cell. Explain why water may move into **cell E** during photosynthesis.

With more glucose in the cell, water will enter by osmosis **(1 mark)**. [1 mark]

iii) The stoma opens as **cell E** takes in water and bends. Explain why **cell E** bends as it takes in water.

The cell fills up with water (becomes turgid) **(1 mark)** *and starts to swell. The thick cell wall on the inside can't stretch, the thinner cell wall on the outside can* **(1 mark)**. *This makes the cell bend.* [2 marks]

How to score full marks

a) The experiment is **investigating transpiration by measuring water loss from the plant**. If Keith didn't cover the soil in plastic film, water would be lost from there as well.

Don't worry if you come across questions that are about experiments you haven't done. You will be given enough information in the question to allow you to answer if you understand the subject, in this case transpiration.

b) You should have learnt this.

c) If you haven't a clue in an exam **you should still attempt an answer**. You might get the first part correct here even if you can't explain it for the second mark. Note that the pores in leaves are called stomata but a single pore is called a stoma.

d) i) **If you've forgotten the formula** for glucose you could work it out by counting how many carbon, hydrogen and oxygen atoms you start with on the left-hand side of the equation.

ii) There are other processes involved in making guard cells take in water, but you are not expected to know about these. You are asked to **'explain why water *may* move'** into the cell, so use knowledge that you do have to suggest an answer.

The fact that the question (ii) is not a whole new question tells you that it is linked to the previous part, which was about glucose, which gives you a clue to the answer.

iii) Look at the diagrams of the guard cells. **You are not usually given unnecessary information in questions**. You should be asking yourself why have you been told about the thick and thin cell walls and expect to use that information somewhere in your answer.

Don't make these mistakes...

Don't say that plants 'breathe'. Oxygen and carbon dioxide enter and leave plants by diffusion. They are not actively moved in and out in the same way as happens in humans when we breathe.

Don't say that plants photosynthesise during the day time and respire at night. They respire all the time.

Chlorophyll does not attract sunlight. Chlorophyll absorbs sunlight.

Don't confuse plant cell walls with cell membranes (see Chapter 1). The job of the cell wall is not 'protection'. The cell wall is inelastic, so if cells are turgid they are firm and this helps to support plants.

Don't describe the minerals that plants take from the soil as 'food', they are not. Plants make their food in photosynthesis.

The waxy cuticle on leaves does not 'prevent water soaking in'. It is water-proof but it's to reduce water loss from the plant, not to stop water getting into the plant.

Don't say that roots support or hold up plants. They do anchor plants in the soil, but it is cell turgor (or sometimes woody tissue) that supports plants.

Plants transport materials through:
- xylem vessels, which carry water and dissolved minerals from the roots to the leaves
- phloem vessels, which carry dissolved food, mainly sucrose, from the leaves to growing and storage regions.

Plants take in minerals from the soil by active transport. Some minerals plants need are:
- nitrates, to make proteins for cell growth
- magnesium, to make chlorophyll.

Farmers and gardeners may add fertilisers to replace minerals lost from the soil.

Plants need food just like animals but they make it for themselves by photosynthesis:

$$\text{carbon dioxide} + \text{water} \xrightarrow[\text{light}]{\text{chlorophyll}} \text{glucose} + \text{oxygen}$$

$$6CO_2 + 6H_2O \longrightarrow C_6H_{12}O_6 + 6O_2$$

Glucose can be used for respiration or converted into other useful substances, such as sucrose (stored in fruit), starch (stored e.g. in seeds) and proteins (for cell growth).

Increasing the carbon dioxide concentration, light intensity and temperature can all increase the rate of photosynthesis until insufficient levels of one of the other factors acts as a limiting factor.

Leaves are adapted for efficient photosynthesis, e.g. palisade cells packed together at top surface, stomata for gas exchange and air spaces to allow gases to circulate.

Plants don't lose water from their leaves to 'remove excess water from the plant'. Some people wrongly compare water loss from leaves with water loss in urine by humans. Plants will only take in as much water from the soil as they need to balance water losses from the leaves.

Plant hormones (plant growth regulators) control many of the ways plants grow and develop.

Plant tropisms are directional growth responses, for example:
- the growth of shoots towards light (phototropism)
- the growth of roots downwards (geotropism).

These tropisms are controlled by the plant hormone auxin.

Many plant hormones are used commercially, for example:
- as selective weedkillers
- as rooting powder
- to delay ripening in soft fruit to prevent damage during transport.

Water enters root hair cells by osmosis and moves to the leaves where it evaporates through the stomata. This is the transpiration stream.

Transpiration is useful because it cools the plant and brings up water and minerals to the leaves. However, if too much water is lost a plant can wilt. To prevent too much water loss leaves have:
- a waxy cuticle
- stomata on the underside where it is cooler
- the ability to close stomata.

Healthy plant cells contain enough water to make them firm (turgid). If water is lost they may become less firm (flaccid). If they lose so much water that the cell membrane peels away from the cell wall, they become plasmolysed.

Plants can grow to enormous heights

Examiner's Hints
(a) You should have learnt this.
(b) Don't describe what has happened, explain it. Use information from the question as well as your own knowledge.

Look at the diagram. It shows a house plant that has been left next to a window.

A plant hormone is making the plant grow towards the window.

(a) Write down the name of this hormone.

_____ [1 mark]

(b) A student decides to investigate this type of growth.

He puts three seedlings next to the window.

The tip of one seedling has been covered in metal foil.

The tip of another seedling has been cut off.

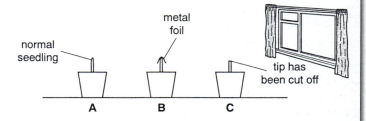

Look at the diagrams below.
They show the seedlings after two days.

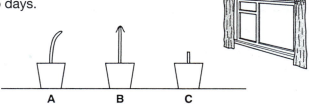

Explain the results shown by seedlings **B** and **C**.
Use ideas about plant hormones in your answers.

(i) Seedling **B**:

_____

_____ [2 marks]

(ii) Seedling **C**:

_____ [1 mark]

Examiner's Hints
(a) Which factor, if increased, will increase the rate of photosynthesis?
(b) Use information from the graph, or your own knowledge.
(c) Think about what the burner will produce as well as what must happen to increase the yield.
(d) Think about what will decide whether he makes more money.

The graph shows the effect of increasing different factors on the rate of photosynthesis.

(a) What factor is limiting the rate of photosynthesis between X and Y?

_____ [1 mark]

(b) Complete the table, by ticking the correct box for each factor, to show the conditions which will produce the highest rate of photosynthesis.

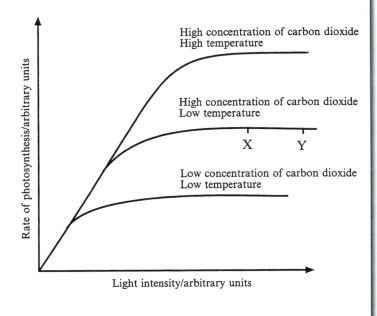

Factor	Level	
	High	**Low**
Light		
Temperature		
Concentration of carbon dioxide		

[1 mark]

A market gardener is trying to find a way to increase his yield of lettuce so that he can make more money. His friend suggests that he should use a paraffin burner in his greenhouse.

(c) Explain how this will increase the yield of lettuce.

_____

_____ [2 marks]

(d) What economic factors must the market gardener consider before he decides if he should carry out this suggestion?

_____

_____ [2 marks]

Answers are given on p.112

5 Ecology and the Environment

Exam Question and Answer

1) a) The diagram shows a simple food web.

    i) Name a primary consumer in the web.

        <u>The tadpoles **(1 mark)**</u>   [1 mark]

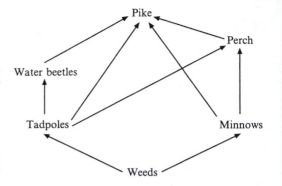

    ii) What process enables producers to provide food for the primary consumers?

        <u>Photosynthesis **(1 mark)**</u>   [1 mark]

    iii) Construct a food chain from the food web consisting of **four** organisms.

        <u>Weeds → tadpoles → perch → pike **(2 marks)**</u>   [2 marks]

    iv) In the food web, which organism would be present in the smallest numbers?

        <u>The pikes **(1 mark)**</u>   [1 mark]

    v) The flow diagram shows the process of eutrophication in the pond.

        Fill in the boxes to complete the diagram.

        [3 marks]

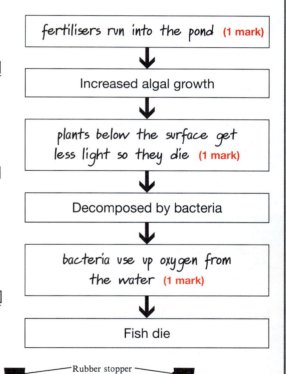

> fertilisers run into the pond **(1 mark)**
>
> ↓
>
> Increased algal growth
>
> ↓
>
> plants below the surface get less light so they die **(1 mark)**
>
> ↓
>
> Decomposed by bacteria
>
> ↓
>
> bacteria use up oxygen from the water **(1 mark)**
>
> ↓
>
> Fish die

b) The diagrams show the apparatus used to investigate the effect of sulphur dioxide, a pollutant, on plants. The seedlings were left for two days.

    i) Why was apparatus B set up?

        <u>To act as a control **(1 mark)**</u>   [1 mark]

    ii) Give **one** way you would expect the seedlings in A to differ from those in B, at the end of the investigation.

        <u>They would not be as tall</u>

        <u>**(1 mark)**.</u>   [1 mark]

    iii) Soot is another pollutant. Explain why soot affects plant growth.

        <u>Soot covers leaves stopping them getting as much light **(1 mark)**</u>

        <u>so they can't photosynthesise as much **(1 mark)**.</u>   [2 marks]

a) i) **Primary consumers eat plants**, so there are two primary consumers here: the tadpoles and the minnows.

ii) All living things need food but **producers (plants) make it for themselves**.

iii) **Food webs are made of different food chains linked together**. There are several food chains here with four organisms. To get both marks you need the correct organisms as well as the arrows.

iv) The answer is the pike because they are at the top of the food web. **The only exceptions to this rule are when animals are feeding on something very much bigger than themselves**, e.g. fleas on a dog.

v) Sometimes you may be asked to describe eutrophication in a paragraph. **Make sure you are clear about all the stages**.

In questions like this where you are asked to write in boxes you should have enough space, **but don't be afraid to carry on outside the box if you need to.**

b) i) A control is to make sure that the results of the investigation are due to the sulphur dioxide, which is what is being investigated here, and not something else, such as the bell jar.

ii) You are asked what you might 'expect' so **any reasonable answer will do**.

iii) This is a different question so expect the soot to have a different effect from the sulphur dioxide.

Don't make these mistakes...

Make sure you write out food chains with the arrows in the correct direction. This is the direction that energy is passing. Remember that arrows mean 'is eaten by'.

If you are asked what a producer is, don't say that it starts off food chains or provides food for animals. These may be true but to be sure of getting the mark say that producers are plants that make their own food (as opposed to getting it by eating other things).

Don't get confused between the different types of chemicals used on crops. Insecticides kill insect pests. Herbicides are weed killers. Fertilisers help plants grow.

Don't get confused between the types of bacteria in the nitrogen cycle: nitrogen fixing, nitrifying and denitrifying. Make sure you know their different roles.

Don't get confused between global warming, acid rain and ozone damage. They are not the same things.

Don't simply say that excessive use of fertilisers can harm rivers and streams because they kill animals and plants. The fertilisers can cause excessive growth of plants such as algae (eutrophication) that eventually die and rot. This uses up oxygen in the water, which kills fish and other animals.

Don't get confused between predators (which hunt other animals) and prey (which are hunted).

Key points to remember

Ecology is the study of the relationships between living things and their environment.

Food chains and webs show how energy passes between living things. Food chains always start with producers, which are green plants that make food by photosynthesis using light energy. Consumers are animals that get food by eating either plants (primary consumers) or animals (secondary or tertiary consumers).

Food chains and webs can also be shown as pyramids of numbers which show the relative numbers at each stage (trophic level).

The raw materials that living things use and are made up of are continually being recycled. Two examples of this are the carbon cycle and the nitrogen cycle.

The main processes in the carbon cycle are:
- respiration of animals and plants, releasing carbon dioxide into the atmosphere
- photosynthesis by plants, removing carbon dioxide from the atmosphere.

The main processes in the nitrogen cycle are:
- nitrogen-fixation, converting nitrogen from the atmosphere into nitrates that can be used by plants
- nitrification, converting the remains of dead and waste material to nitrates
- denitrification, converting nitrates to nitrogen.

These processes involve different types of bacteria living in the soil.

The population sizes of animals and plants usually stay fairly constant, although they may sometimes go up and down from one year to the next. Even though animals and plants continue to reproduce, the population sizes do not continue to increase because of:
- competition for limited resources, e.g. plants may compete for light or water, or animals may compete for food or shelter
- plants or animals being eaten by other animals
- other factors, such as disease or poor weather.

Energy is lost from food chains in several ways, for example:
- respiration (given out as heat)
- egestion
- excretion.

Only a small proportion of the energy at one level of a food chain passes on to the next. This is the reason why pyramids of biomass always show least biomass for the animals at the top of the food chain.

Pyramids of number usually show this pattern as well, although there can be exceptions.

The human population continues to increase, so more resources are being used up and more waste (pollution) is being produced.

Burning of fossil fuels produces carbon dioxide, which is one of the gases contributing to the greenhouse effect which may cause global warming.

Burning fossil fuels also releases gases such as sulphur dioxide that cause acid rain.

Increased amounts of CFC gases are damaging the protective ozone layer in the atmosphere.

To increase crop yields many farmers use chemicals such as insecticides, herbicides and fertilisers which can harm the environment.

Questions to try

(a) Look at the diagram which shows part of the carbon cycle.

What are the **processes** labelled **X**, **Y** and **Z**?

Write your answers in the boxes.

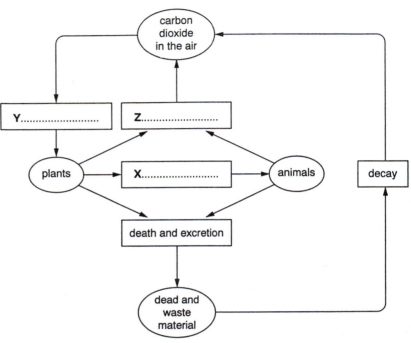

[3 marks]

(b) Nitrifying bacteria and denitrifying bacteria do important jobs in the nitrogen cycle.
What does each type of bacteria do in the nitrogen cycle?

  (i) Nitrifying bacteria: _____

    _____ [1 mark]

  (ii) Denitrifying bacteria: _____

    _____ [1 mark]

Some students do a survey of the animals living in some oak trees. They write down their results in a table.

Feeding level	Animals and plants	Numbers
producers	oak trees	5
primary consumers	insects, squirrels and birds	1017
secondary consumers	spiders and birds	86
tertiary consumers	kestrel	2

(a) Each student uses the results to draw a pyramid of numbers. Look at the pyramids.

Which is the best pyramid to show their results?

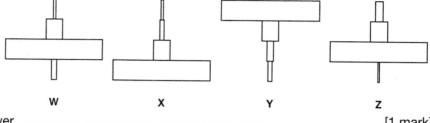

W   X   Y   Z

Choose from **W X Y Z**   Answer _____   [1 mark]

(b) Some of the animals are called **secondary consumers**.

Write about why they are called secondary consumers.

_____

_____ [2 marks]

(c) The students think that a pyramid of biomass is a better way of showing their results.

They will need some more information.

(i) What extra information will they need? _____ [1 mark]

The diagram shows the pyramid of biomass.

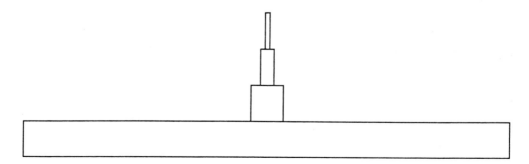

(ii) Explain why the pyramid becomes smaller as it gets to the top.

_____

_____ [2 marks]

Answers are given on p.113

6 Genetics and Evolution

Exam Question and Answer

1) Hassan and Rita both have brown eyes.
Their daughter Melissa has blue eyes.
Human eye colour is decided by genes.
Brown eye colour is dominant to blue eye colour.

a) Explain how Melissa has inherited blue eyes, even though
her parents have brown eyes.

Use these symbols: **B** for brown, **b** for blue.
You can use a genetic diagram.

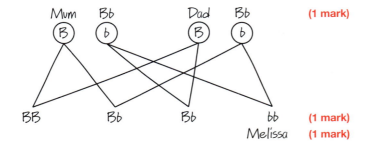

(1 mark)

(1 mark)
(1 mark)

_____

_____

_____

_____ **[3 marks]**

b) Hassan and Rita are both very good at sport.
Melissa may or may not be good at sport.
Explain why.
Use ideas about genes and environment in your answer.

Melissa's parents have genes that make them good at sports **(1 mark)**.
If Melissa has inherited these genes **(1 mark)** then she will probably be good
at sports as well. However even if she has inherited these genes, she
might not be particularly good if she doesn't exercise **(1 mark)** or eat the
right kinds of foods **(1 mark)**.   **[4 marks]**

How to score full marks

a) Although you could give a written explanation **it is probably easiest to use a genetic diagram**. To gain all 3 of the available marks you need to:

- clearly show the parents' genotypes
- clearly show that a **b** is passed on from each parent
- show Melissa as **bb**.

To gain full marks in this question, you don't need to show all the possible children but it is still a good habit to get into. You will not be penalised for doing so and in other questions you may need to show all the children to gain full marks.

b) This question asks you to 'explain' and to 'use ideas about genes and environment'. **As it is worth 4 marks, you know that the examiner will be looking for 4 separate correct points in your answer.**

Other points about genes that would also have been worth 1 mark include:

- Melissa may inherit other genes that would make her poor at sport
- sporting ability may be affected by lots of genes.

Other environmental factors (worth 1 mark each) that can affect sporting ability include:

- motivation
- drugs
- illness or accidents.

There are more than 4 points that you could make. Remember that **time-keeping is important in exams** and once you are sure you have made enough points to get all the marks, move on to the next question. However **it is worth giving a few more than 4 points just in case the examiner decides that not all your points are worth a mark each, but you will lose marks if something you write is incorrect or contradicts another part of your answers.**

Don't make these mistakes...

Remember that people only inherit genes from their parents. They do not inherit features from brothers and sisters or from grandparents or anyone else. They may have some of their genes in common but this is not the same thing.

When you draw genetic diagrams make sure that the sex cells contain only one copy (allele) of each gene whereas parents and children have two. Make sure you have not missed out or repeated possible children.

Don't get confused between mitosis and meiosis. Be careful with the spellings. Poor spelling may not always lose you marks, for example if you incorrectly spell photosynthesis it will probably still be recognisable by the examiner, but if you give an incorrect spelling of mitosis or meiosis it could easily be confused with the other and so the examiner may not be clear which you mean.

Don't get confused between genes and alleles. Alleles are different versions of the same gene. In the example opposite, B and b are different alleles of a gene for eye colour.

In natural selection questions don't make the mistake of suggesting simply that the environment 'makes' the animals or plants change. The environment does not cause the mutations (genetic changes) involved.

The chances of a baby being a boy or a girl are always equal. It does not depend on which sex children a couple already have.

This spider plant is reproducing asexually, producing genetically identical offspring (clones)

Your sex is controlled by the sex chromosomes. In humans, males have an X and a Y chromosome (XY) and females have two X chromosomes (XX). How the sex chromosomes are inherited can be shown in a genetic diagram. There is an equal chance of a baby being a boy or a girl.

Genes are instructions telling cells how to grow and work. Genes are carried by chromosomes in the nucleus of each cell. Humans have 46 chromosomes in each cell except for the sex cells (eggs and sperm) which have 23 (half of 46). Genes are made of a chemical called DNA.

Genes may have dominant or recessive versions. These are known as alleles.

Genetic engineering involves taking the genes from one organism and putting them into the cells of another to give it new abilities or features. The features of animals and plants can also be altered by selective breeding (artificial selection). Make sure you know all the steps in this process.

Practise drawing genetic diagrams. It's usually up to you whether you draw them like the example on page 34 or use a Punnett square. Make sure it is clear which are the parents, the sex cells and the children by labelling them. Genetic diagrams show the probabilities of different types of children.

Cells grow by dividing in half (cell division). Normal cell growth produces cells that are genetically identical and is called mitosis. The other type of cell division is meiosis and only occurs when sex cells are formed. The new sex cells produced only have half the number of chromosomes as normal cells and are genetically different to each other.

People are different to each other (variation) because they have different genes passed on (inherited) from their parents and have had different environmental influences. Environmental influences could include diet, climate, activities, upbringing and many more.

Living things can evolve by natural selection. Make sure you know the steps in this process.

Examiner's Hints
(a) Think about genetic and environmental factors.
(b) Use the information given in the question. Remember that all cell nuclei contain the same number of chromosomes except for the sex cells.
(c) Look at the model answer on page 34. Remember to show **clearly** how the X and Y chromosomes are passed from the parents to the sex cells to the children. Show which sex each child is.

(a) Humans reproduce by sexual reproduction.

Human babies do not grow up to look exactly like either of their parents.

Why don't babies grow up to look exactly like either of their parents?

Suggest **two** reasons.

1 _____

_____

2 _____

_____ [2 marks]

(b) In a human skin cell there are 46 chromosomes.

How many chromosomes would you find in a human liver cell? _____

How many chromosomes would you find in a human sperm cell? _____ [2 marks]

(c) The sex of a person is decided by their sex chromosomes.

Females are **XX** and males are **XY**.

Mr and Mrs Brown have two children, Sarah and James.

Draw a genetic diagram to show how Mr and Mrs Brown had a girl and a boy.

Parents:　　　　　Mrs Brown　　　　　　　　　　Mr Brown
　　　　　　　　　　XX　　　　　　　　　　　　　　**XY**

sex cells:

children:

[2 marks]

Questions to try

Examiner's Hints

(a) The word 'suggest' can mean that you are not necessarily expected to have learnt this. Use your imagination.

(b) Think about what fossils **do** show and so work out what they **don't** show.

(c) Remember the steps involved in natural selection are the same whatever the example.

Some pupils are studying evolution. They are looking at a display in a museum.

Humans might have evolved from ape-like animals. This might have happened in stages.

We cannot be sure because we do not have fossils for every stage.

ape-like animal   human

(a) Why don't we have fossils for every stage of evolution?

Suggest **two** reasons.

1 _____

_____

2 _____

_____ [2 marks]

(b) Fossils cannot show us exactly what the ape-like animals looked like. Why not?

_____

_____ [1 mark]

(c) Many people think that humans have evolved from ape-like animals.
One way we may have changed is that our brains have become larger.
Explain how brains may have become larger as humans evolved.
Use the theory of natural selection in your answer.

_____

_____

_____

_____

_____ [4 marks]

Answers are given on p.114

Exam Question and Answer

1) Read the following information about the blast furnace.

Iron is an important metal. It is used to make a variety of things including bridges, cars and cutlery. Iron is made from iron ores.

If the iron ores in the table are heated strongly in air they will form iron(III) oxide.

Iron is manufactured in a blast furnace. Iron ore, limestone and coke are added to the top of the furnace. Hot air is blown in near the bottom of the furnace. The temperature inside the furnace is above $1000\,°C$. Molten iron and molten slag are collected from the bottom of the furnace.

Many chemical reactions take place in the blast furnace.

Coke contains the element carbon. In the hot temperatures of the blast furnace carbon reacts with oxygen, O_2, to form carbon dioxide, CO_2. Carbon dioxide is then reduced by reaction with further carbon to give the gas carbon monoxide, CO.

Ore of iron	Formula
Haematite	Fe_2O_3
Limonite	$2Fe_2O_3.3H_2O$
Magnetite	Fe_3O_4
Siderite	$FeCO_3$
Iron pyrites	FeS_2

Carbon monoxide reduces iron(III) oxide to make iron and carbon dioxide. The carbon dioxide made either escapes from the top of the blast furnace or reacts with more carbon.

In the high temperature of the blast furnace limestone, $CaCO_3$, decomposes to form calcium oxide and carbon dioxide. The calcium oxide formed in this reaction is used to remove any sand that is present in the iron ore.

a) Look at the table of ores.

   i) Which ore contains three elements chemically combined?

   $FeCO_3$ **(1 mark)** [1 mark]

   ii) Which ore has a formula that contains 19 atoms?

   Limonite **(1 mark)** [1 mark]

   iii) When iron pyrites is heated in air it reacts with oxygen. Sulphur dioxide and iron(III) oxide are the products. Write down the word equation for this reaction.

   iron pyrites + oxygen → sulphur dioxide + iron(III) oxide **(1 mark)** [1 mark]

b) **i)** Write a balanced symbol equation to show the reaction of carbon with oxygen.

   $C + O_2 \rightarrow CO_2$ **(1 mark)** [1 mark]

   ii) Write a balanced symbol equation to show the reaction of carbon with carbon dioxide.

   $C + CO_2 \rightarrow 2CO$ **(2 marks)** [2 marks]

   iii) Write a balanced symbol equation for the reaction of iron(III) oxide with carbon.

   $Fe_2O_3 + 3CO \rightarrow 2Fe + 3CO_2$ **(2 marks)** [2 marks]

How to score full marks

a) **i)** **You must be able to interpret chemical formulae.** The number of elements is shown by the different atomic symbols. Remember you can use the copy of the periodic table provided to check up on the atomic symbols. **You can answer with the formula or the name of the ore because the question did not specify.** If you use the formula make certain that you copy it correctly. **Examiners accept the name of a chemical spelt incorrectly but not an incorrect formula.**

ii) **You obtain the number of atoms in a formula by adding up the numbers of each different type of atom.** You do not always meet formulae with numbers in front or with a dot. The dot separates two formulae and you must add the number of atoms in each formula together. The number in front of each formula means the same as in a balanced symbol equation.

iii) **A word equation must have the reactants on the left and the products on the right hand side of the equation.** Don't include heat in the equation. You can use an arrow or an equals sign in the equation.

b) With a symbol equation **look to see if the formulae are given in the question.** The question is much easier if the formulae are given. **You are expected to know the formulae of some simple substances.** These include:

- the common acids and alkalis
- the oxides of common metals and non-metals
- the hydroxides, carbonates, chlorides, nitrates and sulphates of common metals.

Remember to put the formulae in the correct part of the equation. **Often there will be 1 mark for just putting the formulae of the reactants on the left and the products on the right.** Don't change these formulae. Only put numbers in front of them to balance them.

Don't make these mistakes...

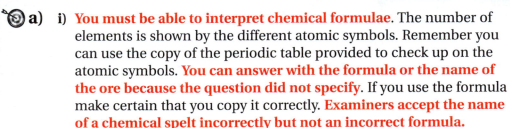

It will be impossible to balance an equation unless you have the correct formulae. So make sure you copy the formulae down carefully.

Don't change a formula when balancing an equation. Calcium oxide is CaO and not CaO_2 even if CaO_2 might help you balance an equation.

Don't write the name of a substance when the question asks for a formula or write the formula of a substance when it asks for a name.

Don't put anything in an equation other than the reactants and the products. Don't put heat into an equation or the name of a catalyst. You will sometimes see the conditions put over the arrow in an equation but it is safer for you not to do this.

Don't use an arrow when the reaction is reversible and there is an equilibrium. In reversible reactions use the symbol ⇌

Don't just balance the atoms in an ionic equation. You must balance the atoms and the charge.

Don't use a capital letter in a formula when it should be a lower case letter. The symbol for iron is Fe and not FE.

Key points to remember

Practise working out the formulae of compounds. You can't balance a symbol equation unless you have the correct formula.

You can work out the formulae of ionic compounds from the formulae of the cations and anions present.

An empirical formula shows the simplest ratio of each type of atom in a compound. The chemical formula of an ionic compound is really an empirical formula.

A displayed formula shows the number and type of each atom in a particle (normally a molecule) as well as how each atom is bonded to other atoms. Covalent bonds are shown with a line between atoms.

Remember the formulae of the following compounds:
- carbon dioxide, water, hydrogen, oxygen, hydrochloric acid, nitric acid, sulphuric acid and ammonia
- chlorides, nitrates, sulphates, carbonates, hydroxides, and oxides of the metals sodium, potassium, calcium, magnesium, copper(II), iron(II) and zinc.

A molecular formula gives the number and type of each atom in a molecule.

An ionic compound cannot have a molecular formula.

A balanced symbol equation has the formulae of the reactants on the left hand side and the formulae of the products on the right hand side. An arrow separates the reactants and the products. The number of each type of atom on the left hand side of the equation must equal the number of each type of atom on the right hand side.

In a chemical reaction reactants are changed into products.

A word equation has the names of the reactants on the left hand side and the names of the products on the right hand side. An arrow separates the reactants and the products.

There are several types of chemical formulae:
- molecular formulae
- displayed formulae
- empirical formulae.

State symbols can be used to tell you if the substances are:
- aqueous (aq)
- liquid (l)
- solid (s)
- gas (g).

Whether the reaction is exothermic or endothermic can also be indicated in an equation:
- $\Delta H = -$ is exothermic
- $\Delta H = +$ is endothermic.

Practise writing balanced symbol and word equations.

This scanning electron micrograph of a salt crystal shows the cubic lattice of sodium and chloride ions

Look at the following formulae.

Na

Na^+

NaOH

Na_2CO_3

H_2SO_4

HNO_3

NH_3

SO_4^{2-}

Cl^-

NH_4^+

(a) Choose from the list the formula for:

  (i) an ion that contains two elements

  _____ [1 mark]

  (ii) a cation

  _____ [1 mark]

  (iii) sodium carbonate

  _____ [1 mark]

  (iv) sulphuric acid

  _____ [1 mark]

  (v) ammonia

  _____ [1 mark]

(b) Write the formula for ammonium sulphate.

  _____ [1 mark]

Barium, Ba, forms a positive ion with the formula Ba^{2+}. Barium reacts with water according to the following equation.

$Ba(s) + 2H_2O(l) \rightarrow Ba(OH)_2(aq) + H_2(g)$ ΔH = −ve

(a) (i) Write down the word equation for this reaction.

_____ [1 mark]

(ii) Suggest what you would observe when barium is added to water.

_____

_____

_____ [3 marks]

(b) The nitrate ion has the formula NO_3^-.

(i) What is the formula for barium nitrate?

_____ [1 mark]

(ii) What is the formula for barium oxide?

_____ [1 mark]

(iii) What is the formula for barium chloride?

_____ [1 mark]

Answers are given on p.115

Exam Question and Answer

1) This question is about magnesium oxide and how it is formed from atoms of magnesium and oxygen.

a) Look at the table. It gives information about magnesium atoms and oxygen atoms.

Atom	Number of protons	Number of neutrons	Number of electrons	Electron arrangement
Mg	12 **(1 mark)**	12	12 **(1 mark)**	2.8.2
O	8	8	8 **(1 mark)**	2.6 **(1 mark)**

Complete the table. [4 marks]

b) What is the mass number (nucleon number) of the magnesium atom shown in the table?

24 **(1 mark)** [1 mark]

c) Describe and explain the change in electron arrangement that takes place when magnesium oxide is formed from magnesium and oxygen.

A magnesium atom loses two electrons **(1 mark)** so that it has the same electron structure as the noble gas neon, 2.8 **(1 mark)**. An oxygen atom gains two electrons **(1 mark)** so that it has the same electron structure as neon **(1 mark)**. [4 marks]

d) Write a balanced symbol equation for the reaction between magnesium and oxygen.

$2Mg + O_2 \rightarrow 2MgO$ **(2 marks)** [2 marks]

e) **i)** Magnesium oxide has a giant ionic structure. Magnesium oxide has a very high melting point. Explain why.

Very strong ionic bonds **(1 mark)** between the positive magnesium ion and the negative oxide ion means that it needs a large amount of energy to break the bonds **(1 mark)**. [2 marks]

ii) Give one other physical property of solid magnesium oxide.

Does not conduct electricity **(1 mark)**. [1 mark]

How to score full marks

a) The answers to this question rely on **knowledge** and **understanding** of **atomic structure**. You should know that **atoms are neutral** so that the **number of protons always equals the number of electrons**. You ought to remember the electron structure of the first 20 elements in the periodic table, or know how to work them out from a copy of the periodic table.

b) You can't work out the mass number (nucleon number) from a periodic table. **Only the atomic number (proton number) is shown on a periodic table**. Add the number of protons and the number of neutrons given in the table in the question.

c) This question asks you to **describe** and **explain**, so you can be sure that at least 1 mark, if not 2, are for the explanation. As the question is worth 4 marks, you know the examiner will want **4 separate points**. Notice that in the answer 2 marks are for describing electrons being lost or gained and the other 2 for explaining why they are lost or gained.

d) You should have been able to work out the formula of magnesium oxide because one atom of magnesium loses two electrons, which one atom of oxygen gains. You have to remember that **oxygen exists as molecules, O_2**. When writing a balanced symbol equation don't attempt to change the formula once you are certain that you have them right. **Only change the numbers in front of the formulae**.

e) **i)** When a question asks you to **explain**, this often means you have to **link two ideas**. In this case it is **the presence of strong bonds that need lots of energy to break**.

ii) You must try and **interpret** some questions when they seem to need obscure knowledge. This question is not really about magnesium oxide but about **giant ionic structures**. High boiling point would not be given credit because it is too closely linked to melting point.

Don't make these mistakes...

Don't waste time in an exam drawing a 'dot and cross' diagram using a compass.

Don't forget that the periodic table can be used to work out a number of important things:
- atomic symbols and names of elements
- number of electrons in outer shell (group number)
- atomic number
- number of shells (period number).

Don't forget that electrons are gained or lost from atoms to make ions not protons.

Don't get confused between covalent bonds and intermolecular forces. Covalent bonds are between atoms within a molecule, and intermolecular forces are between molecules. Covalent bonds are much stronger than intermolecular forces.

Don't be imprecise when referring to bonds. Make certain that you specify the bonds you are writing about. This is particularly important when explaining the melting point of a substance. For example, do not write because the bonds are weak; state which bonds you are referring to. Examiners expect precision in this context.

Don't draw ionic 'dot and cross' diagrams with shared pairs of electrons. Keep the electron structure of the two ions separate and make certain that you include the charge on the ion. With covalent 'dot and cross' diagrams, make certain you show that the electrons are shared. Use a dot and a cross, or crosses with different colours, to indicate from which atom the electrons come.

Electrons occupy shells or orbits around the nucleus. Make sure you can work out the electron structure of the first 20 elements in the periodic table, e.g. calcium is 2.8.8.2

The atomic (proton) number is the number of protons in an atom. The mass (nucleon) number is the total number of protons and neutrons in an atom. Isotopes are varieties of an element that have the same atomic number but different mass numbers.

A giant ionic structure has positive ions electrostatically attracted to negative ions. Magnesium oxide and sodium chloride have giant ionic structures. A substance with a giant ionic structure will:
● have a high melting point
● not conduct electricity as a solid
● conduct electricity either in solution or as a molten liquid.

A molecule is two or more atoms bonded together by covalent bonds. A covalent bond is a shared pair of electrons. A displayed formula shows the number and type of atoms in a molecule and how they are bonded together.

Simple molecular structures have weak intermolecular forces between molecules. Water and carbon dioxide have simple molecular structures. A substance with a simple molecular structure will
● have a low melting and boiling point
● not conduct electricity.

An ion is a charged atom or group of atoms. Positive ions are called cations and are formed by the loss of electrons. Negative ions are called anions and are formed by the gain of electrons.

A metal and non-metal combine by transferring electrons to form positive ions and negative ions which then attract one another.

A non-metal and another non-metal combine by sharing electrons. A shared pair of electrons holds the atoms together.

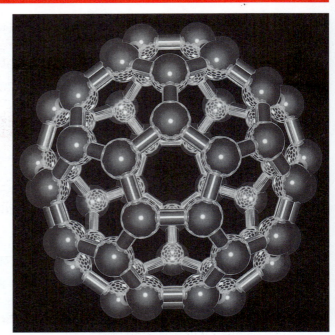

This computer graphic shows a buckyball molecule. The spheres represent carbon atoms, linked by covalent bonds.

An atom has a nucleus, made up of protons and neutrons, surrounded by electrons.

The relative charge and relative mass of an electron, a proton and a neutron are
● electron charge −1 and mass 0.0005 (zero)
● proton charge +1 and mass 1
● neutron charge 0 and mass 1.

All atoms try to gain, lose or share electrons in order to obtain a stable octet of electrons in their outer shell (this will be the electron arrangement of a noble gas). 'Dot and cross' diagrams can be used to describe the bonding in ionic and in covalent compounds. Make sure you can draw 'dot and cross' diagrams.

A giant molecular structure has all of its atoms joined by covalent bonds. A whole crystal may be one molecule. Diamond and graphite have giant molecular structures. A substance with a giant molecular structure will:
● have a very high melting point
● not dissolve in water
● normally not conduct electricity.

Questions to try

Examiner's Hints

(a) The charge on a negative ion tells you how many electrons it has gained from another atom. In the formula of an ionic compound, the charge of the positive ions will cancel out the charge on the negative ions.

(b) When drawing a 'dot and cross' diagram make certain you are sure whether it is ionic or covalent. If it is covalent then electrons must be shared as a pair; if it is ionic then draw the negative ion and the positive ion separately so there is no chance of you sharing electrons.

(c) Remember that some of the physical properties of a solid will depend on the force of attraction between particles; it is strong between ions but weak between molecules.

This question is about sodium, hydrogen and chlorine. Sodium is a metal and hydrogen and chlorine are both non-metals.

(a) Sodium hydride is an ionic compound containing sodium and hydrogen only. The hydride ion has the formula H^-.

(i) Predict the formula for sodium hydride.

_____ [1 mark]

(ii) Explain how a hydride ion is formed from a hydrogen atom.

_____ [1 mark]

(b) Look at the diagram. It shows the electron structure for chlorine.

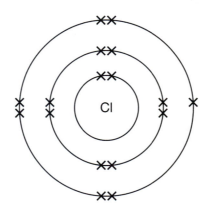

(i) Draw a similar diagram to show the electron structure of sodium.

[1 mark]

(ii) Draw a 'dot and cross' diagram to show
the bonding in sodium chloride.

[3 marks]

(iii) Sodium chloride has a giant ionic structure. Explain why solid sodium chloride
does not conduct electricity but molten sodium chloride will.

_____

_____

_____ [2 marks]

(c) Chlorine and hydrogen react to form hydrogen chloride, HCl.

(i) What type of bonding would you expect between a hydrogen atom and
a chlorine atom? Explain your answer.

_____

_____

_____ [2 marks]

(ii) Draw a 'dot and cross' diagram to show the
bonding in a molecule of hydrogen chloride.

[2 marks]

(iii) Construct the balanced equation for this reaction.

_____ [1 mark]

(iv) Solid hydrogen chloride has a simple molecular structure. Predict **two** physical
properties of solid hydrogen chloride.

_____

_____

_____ [2 marks]

Answers are given on p.116

Exam Question and Answer

1) Crude oil or petroleum is a complex mixture of hydrocarbons that contains hundreds of different compounds.

Crude oil can be separated into fractions by distillation.

Look at the table. It shows different fractions that can be obtained from crude oil.

Fraction	Number of carbon atoms per molecule	Boiling range
Liquefied petroleum gases	1–4	Less than 40°C
Petrol (gasoline)	5–9	40–70°C
Naphtha	8–12	60–210°C
Kerosene (paraffin)	10–14	190–260°C
Gas oil (diesel)	14–20	250–340°C
Lubricating oil	19–30	330–500°C
Bitumen	Over 30	Above 450°C

a) The molecule of a hydrocarbon contains 11 carbon atoms.

Estimate the boiling point of hydrocarbon.

200 °C (1 mark)   [1 mark]

b) What is a hydrocarbon?

A compound that contains only hydrogen and carbon (1 mark)   [1 mark]

c) Distillation of crude oil gives fractions that are still mixtures. It does not give pure single compounds. Suggest why.

There are so many different compounds in crude oil that there is little difference between the boiling points of each compound (1 mark). Fractional distillation only separates mixtures if the components have quite a large difference in boiling points (1 mark).   [2 marks]

d) The boiling temperature of diesel is much higher than that of petrol. Explain why.

Use information from the table and ideas about the forces between molecules.

The larger the hydrocarbon molecule the higher its boiling point. Diesel contains larger hydrocarbon molecules than petrol (1 mark). The intermolecular forces in diesel are much stronger than in petrol (1 mark).   [2 marks]

e) The liquefied petroleum gas fraction contains gaseous alkanes.

Write the molecular formula of the gaseous alkane which has one carbon atom in its molecule.

CH_4 (1 mark)   [1 mark]

a) You must be able to **interpret numerical data**. In the table, two fractions contain hydrocarbon molecules with 11 carbon atoms per molecule. You can estimate the boiling point using the boiling range from both fractions.

b) Some definitions must be very precise otherwise they are wrong. **A hydrocarbon is a compound** and **not a mixture of** hydrogen and carbon. A hydrocarbon **only contains hydrogen and carbon** and no other elements, so an alternative answer could be that it is a substance that contains only carbon and hydrogen chemically bonded together.

c) The word 'suggest' in a question means that you must **work out the answer** and **that there may be more than one possible answer**. You need to link the idea that fractional distillation only separates mixtures where the components have boiling points that differ by at least 10°C, with the knowledge that crude oil is an extremely complex mixture containing thousands of compounds, many of which have the same or very similar boiling points.

d) You must clearly link **the strength of the force between molecules with the boiling range. Don't refer to the strength of covalent bonds** because these are bonds within a molecule. Don't write that the bonds in petrol are stronger than in diesel. There is no indication what bonds are being described: they could be covalent bonds within a molecule or bonds between molecules. When using the word bond or force make sure you define what the force or bond is between. **Make certain you make the link between the size of the molecule and the strength of the intermolecular force**; from the question it must be there to get full marks.

e) You should know that the **general formula of an alkane is C_nH_{2n+2}** and so it is easy to work out the molecular formula when $n = 1$.

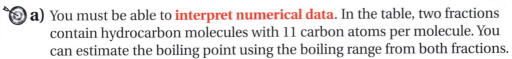

Don't make these mistakes...

Don't confuse a hydrocarbon with a carbohydrate. A carbohydrate contains carbon, hydrogen and oxygen chemically combined.

Don't confuse cracking with polymerisation. Cracking involves the breakdown of large alkane molecules to get smaller hydrocarbons. Polymerisation involves many smaller molecules joining together to form a very long chain molecule.

$\Delta H = +ve$ is not an exothermic reaction. The plus sign does not mean that energy or heat is released it means that the products have more energy than the reactants.

Don't confuse intermolecular forces and intramolecular forces. One molecule is attracted to another molecule by intermolecular force; these are often weak forces of attraction. Intramolecular forces called covalent bonds hold atoms within a molecule together. Covalent bonds are very strong.

Don't confuse the greenhouse effect with ozone depletion. The greenhouse effect is the trapping of energy (IR) radiated from the Earth's surface, caused by greenhouse gases in the atmosphere. Ozone depletion is the loss of ozone molecules from the atmosphere, caused by CFC gases, allowing more dangerous UV light to reach the Earth's surface.

Key points to remember

Crude oil is formed from marine animals and plants that have been compressed underground over millions of years. Crude oil is a mixture of hydrocarbons. Crude oil is separated by distillation in a fractionating column to give fractions that have different boiling ranges.

The bottom of a fractionating column is the hottest not the coolest part of the column. Fractions with low boiling temperatures 'exit' at the top of the fractionating column, whereas fractions with high boiling temperatures 'exit' from the bottom.

You should be able to list and to be able to give uses for some of the major fractions obtained from crude oil, e.g. petrol, diesel, paraffin, liquefied petroleum gases, heating oil, lubricating oil, bitumen and naphtha.

A fuel is a substance that will react or burn in oxygen to release large amounts of useful energy, normally in the form of heat. Combustion or burning is an example of an oxidation. Combustion of fuels gives rise to pollution problems such as the greenhouse effect and acid rain. Nuclear fuels do not burn.

Complete combustion needs a plentiful supply of oxygen (air). Complete combustion of a hydrocarbon fuel gives carbon dioxide and water.

Incomplete combustion takes place when there is a shortage of oxygen. Incomplete combustion of a hydrocarbon fuel gives water and either carbon or carbon monoxide.

Practise writing word and symbol equations for the complete and incomplete combustion of hydrocarbon fuels.

An alkane is a saturated hydrocarbon. The atoms in a molecule of an alkane are bonded by single covalent bonds.

An alkene is an unsaturated hydrocarbon that contains at least one double bond between carbon atoms. Alkenes react by addition.

Practise writing the displayed formulae of hydrocarbons that contain four carbon atoms or less per molecule.

Cracking enables more useful fractions from distillation, e.g. petrol, to be made from less useful fractions, e.g. naphtha alkane. Large alkane molecules can be broken down into smaller alkanes and alkenes. Cracking requires a catalyst and a high temperature.

Polymerisation involves the reaction of many small molecules, monomers, to form a large polymer molecule. Many monomers are alkenes, which are available from cracking and can be used to form addition polymers. Polymerisation requires a high pressure and catalyst.

The energy released by a fuel can be measured by using it to heat water:

energy released = mass of water heated × 4.2 × temperature change

Remember that the mass is the mass of water heated not the mass of the fuel.

An exothermic reaction transfers energy into the surroundings so that the surroundings get warm. A reaction is exothermic because more energy is released during bond making than is absorbed during bond breaking. Bond making is exothermic.

An endothermic reaction absorbs energy from the surroundings so that the surroundings get colder. Bond breaking is endothermic.

Renewable fuels can be made in a short space of time so that the fuel will not run out.

Non-renewable fuels take millions of years to form but are used up very quickly. Non-renewable fuels cannot be used again and will eventually run out.

The carbon cycle helps to maintain the composition of the atmosphere. Photosynthesis adds oxygen and removes carbon dioxide, whereas combustion and respiration add carbon dioxide and remove oxygen.

During boiling intermolecular forces are broken. The intermolecular forces between large hydrocarbon molecules are stronger than those between smaller hydrocarbon molecules. Hydrocarbons with large molecules have a higher boiling temperature than those with smaller molecules. Remember that covalent bonds are not broken during boiling.

Ethanol, C_2H_5OH, and methanol, CH_3OH are two liquid fuels.

(a) Ethanol can be made by fermentation of a glucose solution.

Ethanol produced in this way is renewable.

Explain why.

_____

_____ [2 marks]

(b) (i) What is a hydrocarbon?

_____

_____ [2 marks]

(ii) Explain why methanol is not a hydrocarbon.

_____ [1 mark]

(c) Aroon wants to find out whether one gram of ethanol or one gram of methanol gives out more energy when it is completely combusted.

Describe an experiment to show how she could do this.

Include the apparatus Aroon uses, the measurements she makes and how she will decide which fuel gives out more energy.

_____

_____

_____

_____

_____ [6 marks]

(d) Look at the equation.

It shows the complete combustion of methanol.

$$H-\underset{\underset{H}{|}}{\overset{\overset{H}{|}}{C}}-O-H \;+\; \tfrac{3}{2}\,\overset{O}{\underset{O}{\parallel}} \longrightarrow \underset{O}{\overset{O}{\parallel}}C \;+\; 2\,O\!\!<\!\!\begin{array}{c}H\\[2pt]H\end{array}$$

(i) Write the names of **two** products made during the incomplete combustion of methanol.

_____

_____ [2 marks]

(ii) The complete combustion of methanol is exothermic.

Explain why.

Use ideas about bond breaking and bond making.

_____

_____ [3 marks]

Examiner's Hint

● Write about both processes in your answer. Remember cracking involves shortening the carbon chain and polymerisation lengthening it.

Fractional distillation of crude oil produces useful fractions and less useful fractions. The less useful fractions are cracked to form alkenes and more useful hydrocarbons.

Alkenes can be polymerised to make plastics.

Distinguish between the processes of **cracking** and **polymerisation**.

_____

_____

_____

_____

_____ [6 marks]

Answers are given on p.117

Exam Question and Answer

1) The diagram opposite shows part of a rock face.

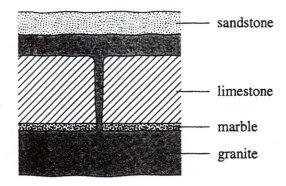

a) **i)** Name **one** sedimentary rock.

Limestone **(1 mark)**   [1 mark]

ii) Explain how sedimentary rocks are formed.

Very small grains of rock are compacted together **(1 mark)** under pressure from other layers. The grains are then cemented together **(1 mark)** to form a rock.

[2 marks]

b) **i)** What **type** of rock is marble?

Metamorphic **(1 mark)**   [1 mark]

ii) Explain how it was formed from limestone.

Limestone was subjected to a high temperature but it did not melt **(1 mark)** and high pressure **(1 mark)** where it changed form to make marble.

[2 marks]

c) **i)** In the above rock face which was formed first, the limestone or the granite?

Limestone **(1 mark)**   [1 mark]

ii) Explain your answer.

Normally the lower the layer of rock the older it is, however this only applies if the rocks are all sedimentary **(1 mark)**. Granite is an igneous rock formed when molten magma is cooled down. Some of the molten magma was able to force its way through the layer of limestone rock **(1 mark)**. This means that the limestone must be older than the granite.

[2 marks]

How to score full marks

a)

i) Although it was not stated in the question **you must select the answer from the diagram**. Another sedimentary rock in the diagram is sandstone.

ii) The answer given states the two key processes that occur during the formation of sedimentary rock. **Compaction and cementation occur whenever a sedimentary rock is formed whether it is under water or not.** The mark allocation shows that only two key points are needed. If the question was worth more marks, you would have to write about erosion, weathering or transportation to get full marks.

b)

i) You need to be able to **recognise one or two common examples of each type of rock**. **Sandstone** and **limestone** are sedimentary rocks. **Marble** and **slate** are metamorphic rocks. **Granite**, **basalt** and **pumice** are igneous rocks.

ii) This question asks you to **explain** and you should **use the information given on the diagram to explain** rather than just describe. The presence of marble next to the granite indicates that the marble was made when hot magma was cooling down, and so a good answer **links the presence of the granite with the formation of marble**.

c) You need to **refer to the increasing age of sedimentary rock as you get deeper**, and say that **any rock which intrudes into a layer must be younger than any rock in the layer**. Granite is an igneous rock so it was formed at a different time and in a different way to the layers of sedimentary rock. The answer only needs 2 points to score full marks, so don't waste time giving many more.

Don't make these mistakes...

Don't get confused between magma and lava. Magma is molten rock beneath the surface of the Earth, and lava is molten rock that comes out of a volcano.

Don't get confused about the electric charge on ions and electrodes. Cations are positive but the cathode is negative. Anions are negative but the anode is the positive electrode.

Don't confuse the terms mineral and ore. An ore is a mixture of rock and a mineral that is the source of a metal or a metal compound. A mineral is a chemical compound found in the ground that contains a metal.

Only solutions that contain charged particles that can move can conduct electricity. Just because a substance contains ions does not mean it will conduct electricity. The substance must be melted or dissolved in water so that the ions are free to move.

In a chemistry question the word metal refers to an element not an alloy, which is a mixture of elements.

Don't think that tectonic plates are part of the crust. Tectonic plates are part of the lithosphere, which is the rigid layer at the surface of the Earth that consists of the crust and the upper part of the mantle.

Don't think that sedimentary rock has to be formed under water. Although some sedimentary rock such as limestone is formed under water, others are not.

Don't write ions without showing the charge as a superscript.

Don't write the formula of a compound unless it is specifically asked for in the question. Examiners will allow the names of chemicals to be incorrect providing there is no confusion, but a formula has to be completely correct. So potassium cloride would be an acceptable form of potassium chloride but KCl_2 would not (the actual formula is KCl).

Key points to remember

You should be able to draw and interpret simple rock cycles involving sedimentary rock, metamorphic rock, igneous rock and molten rock in the mantle.

Igneous rocks are formed when molten magma or lava cools down. Igneous rock can also be formed from volcanic ash. Igneous rock often has interlocking crystals or a glassy appearance. They often intrude into another rock. Any rock produced by an intrusion will be younger than the surrounding rocks.

The Earth is a sphere with a:
- thin rocky crust
- mantle of mostly solid rock
- iron core.

Metamorphic rocks are formed when igneous or sedimentary rocks are changed by temperature and/or pressure. Metamorphic rock is formed by recrystallisation of grains (without melting) and any fossils present can be distorted or destroyed.

The reactivity series can be used to make predictions about:
- displacement reactions
- method of extraction.

Learn and be able to define the following terms:
- electrolysis
- cathode and anode
- electrolyte
- cation and anion.

Practise writing word and symbol equations involving all the metals on your syllabus, including the reaction of metals with air, water and acids.

The lithosphere is the (relatively) cold rigid outer part of the Earth that includes the crust and the outer part of the mantle. The lithosphere is made of a number of large interlocking tectonic plates. The tectonic plates are less dense than the rocks underneath them. There are two types of continental plates:
- oceanic
- continental.

Unreactive metals, e.g. gold and silver, are found uncombined in nature and only need to be separated from rocks using physical methods.

Sedimentary rocks are formed from sediments when they are compacted and/or cemented. Sedimentary rocks are found in layers and the bottom layer is most likely to be the oldest layer. Sedimentary rocks may contain fossils and the ages of the fossils found can indicate the age of the rock (the younger the fossils the younger the rock).

The reactive series of metals is: potassium, sodium, lithium, calcium, magnesium, aluminium, zinc, iron, lead, copper, silver and gold (most reactive to the least reactive).

The theory of plate tectonics explains:
- continental drift
- the rock cycle, including subduction and the formation of metamorphic rock
- folding and mountain building
- earthquakes and volcanoes.

Rocks and minerals provide the raw materials for making metals. Extraction is the name given to the chemical processes involved in obtaining a metal from its ore. Extraction of metals involves reduction (the gain of electrons or the loss of oxygen).

Reactive metals, e.g. sodium and aluminium, are extracted using electrolysis. Aluminium is extracted by the electrolysis of molten aluminium oxide. Moderately reactive metals, e.g. copper and iron, are extracted by heating the ore with carbon or carbon dioxide. Iron is manufactured in the blast furnace.

Questions to try

Examiner's Hints

(a) Use the information in the text about how energy can be saved.

(c) Think about the particles present in the electrolyte.

(e) Remember, in any equation write down what you start with on the left and what you end up with on the right. You will need to use the symbol for an electron, e⁻, to balance the charge in this equation.

Aluminium is manufactured by the electrolysis of molten aluminium oxide.

Aluminium oxide is dissolved in molten cryolite (another compound of aluminium) at a temperature of over 900 °C. If the molten aluminium oxide on its own were used then a temperature of about 2000 °C would have to be used.

The electrolyte contains the following particles. Al^{3+} F^- O^{2-}

(a) Explain why it is a good idea to electrolyse a mixture of molten cryolite and aluminium oxide rather than molten aluminium oxide on its own.

_____

_____

_____ [2 marks]

(b) The electrolytic manufacture of aluminium uses an anode made from carbon. What is the cathode made from?

_____ [1 mark]

(c) Explain how the electrolyte conducts electricity.

_____

_____

_____ [2 marks]

(d) Oxide ions react at the anode. What gas is made at the anode?

_____ [1 mark]

(e) Aluminium ions react at the cathode. Write down the electrode reaction that happens at the cathode and explain why it is a reduction.

_____

_____

_____ [3 marks]

Describe how the theory of plate tectonics can explain:

mountain building and why the continents of Africa and South America are slowly moving apart.

Include in your answer how tectonic plates move.

_____

_____

_____

_____

_____

_____

_____

_____

_____

_____

_____

_____ [6 marks]

Answers are given on p.118

Exam Question and Answer

1) Look at the flow chart. It shows the main stages in the manufacture of ammonia by the Haber process.

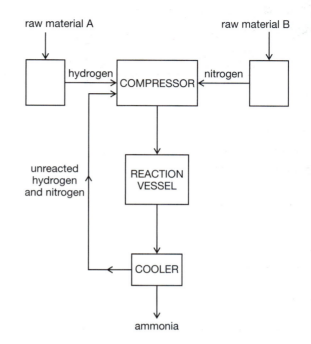

a) i) Write down the name of raw material A.

Crude oil **(1 mark)** [1 mark]

ii) Write down the name of raw material B.

Air **(1 mark)** [1 mark]

b) The reaction vessel contains an iron catalyst. Explain why.

So that the reaction will go faster. **(1 mark)** [1 mark]

c) Write down the balanced equation for the reversible reaction to make ammonia.

$N_2 + 3H_2 \rightleftharpoons 2NH_3$ **(2 marks)** [2 marks]

d) Look at the table. It gives some information about the effect of pressure and of temperature on the percentage conversion of nitrogen and hydrogen into ammonia.

Typical conditions used in the Haber process are 250 atmospheres and 450 °C. Explain why. Use ideas about rate of reaction and equilibria.

Temperature in °C	Pressure in atmospheres	Percentage conversion
250	200	75
250	1000	96
1000	1	0.01
1000	1000	1

If the temperature is too low then the rate of reaction is too slow **(1 mark)**. If the temperature is too high the percentage of hydrogen and nitrogen converted is too low **(1 mark)**. This means that a temperature of 450°C is a compromise, the reaction is quite fast and there is a reasonable percentage conversion **(1 mark)**. A very high pressure of 1000 atmospheres gives a rapid rate of reaction **(1 mark)** and it gives a very high percentage conversion **(1 mark)**. If the pressure were too low then the percentage conversion would be too low **(1 mark)** so a compromise pressure of 250 atmospheres is used **(1 mark)**. At 200 atmospheres and 450°C not all the hydrogen and nitrogen is converted but in the industrial process the unconverted gases are recycled **(1 mark)**. [8 marks]

How to score full marks

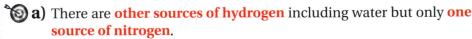 **a)** There are **other sources of hydrogen** including water but only **one source of nitrogen**.

 **b)** Sometimes it is the obvious that you need to include. **An even better answer would refer to the lowering of the activation energy which will increase the rate of reaction**.

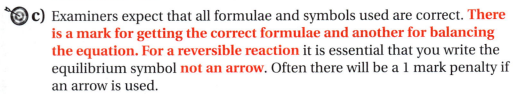 **c)** Examiners expect that all formulae and symbols used are correct. **There is a mark for getting the correct formulae and another for balancing the equation. For a reversible reaction** it is essential that you write the equilibrium symbol **not an arrow**. Often there will be a 1 mark penalty if an arrow is used.

d) **In your exam you will be faced with some questions that need extended answers and others that involve information you must analyse**. This question has both. Examiners expect you to **use the data in the table first of all** and **then apply your knowledge and understanding** later on. Use each piece of data in the table and make a sensible comment about the data. For example, look at the first two lines of the table. These show that as the pressure increases (temperature constant) the conversion of nitrogen and hydrogen increases.

The answer given has not referred to economic issues. **You could get additional marks** by referring to the expense of building and maintaining a factory that uses 1000 atmospheres, or by mentioning that the catalyst allows for a much lower temperature to be used, so saving on energy (and increasing the percentage conversion). You could also refer to Le Chatelier's principle to get more marks.

Don't make these mistakes...

Don't confuse the role of a catalyst. Remember, a catalyst only affects the rate of the reaction not the position of equilibrium.

Don't ignore the data given in a table. Data in a table are there to help you. Make full use of the data in your explanations.

Don't forget to refer to collisions and collision frequency when explaining the effect of changing conditions on the rate of reaction.

Don't confuse rate of reaction with the position of equilibrium. Remember that even with the position of equilibrium on the right hand side the reaction may be very slow.

Don't confuse the effect of rate of reaction and the mass of product formed. Remember, the mass of product formed depends on the mass of reactants, not how fast the reaction is.

Don't miss the obvious points because they are too simple. Your answer provides the only opportunity for you to demonstrate your knowledge and understanding to the examiner.

Don't confuse the effect of temperature on an enzyme. Remember, enzymes are denatured or destroyed but not killed.

Don't make a series of isolated points if a question is asking for a piece of continuous writing. Try to link points together in a logical way.

Don't forget to include the units when referring to physical quantities. A pressure of 200 could mean atmospheres or Pascals: so include the unit. Sometimes there will be a mark for including the unit.

Don't confuse the command words such as describe or explain given in a question. Explain means that you have to demonstrate your knowledge and understanding.

Key points to remember

Increasing the temperature shifts the position of equilibrium to the endothermic reaction.

A catalyst has no effect on the position of equilibrium.

Increasing the pressure shifts the equilibrium to the side of the equation with the least number of moles of gas.

Equilibria

Chemical processes involving equilibria want the highest percentage conversion, the highest rate of reaction at the cheapest cost, so compromise conditions are chosen.

Reversible reactions can occur in both directions.

Decomposition is when one substance is broken down into two or more simpler substances.

Redox involves oxidation, the loss of electrons or gain of oxygen, and reduction, the gain of electrons or loss of oxygen.

Chemical reactions

Precipitation is where two solutions react to make an insoluble solid.

Neutralisation is where a base and an acid make a salt and water.

Combustion is a reaction in which a fuel reacts with oxygen to release useful energy (heat).

The rate increases as the temperature increases due to an increase in the number of effective or successful collisions.

Rate of reaction

An increase in concentration will increase the rate of reaction because the particles are more crowded so collision frequency increases.

A catalyst normally increases the rate of a reaction because it lowers the activation energy. Enzymes are biological catalysts.

An increase in surface area will increase the rate of reaction because there is an increase in the collision frequency.

The rate increases in gas phase reactions as the pressure increases due to an increase in the collision frequency.

61

Examiner's Hints

(b) (i) Think about how temperature changes the rate of a reaction.

(b) (ii) Compare the mass of zinc used in the different experiments.

(c) Think about how reactions take place and how the arrangement of particles varies from concentrated to dilute acid.

(d) Think about the difference in surface area between zinc lumps and zinc powder.

(e) Use the information given in the question. Remember that a catalyst speeds up a reaction but is chemically unchanged at the end of the reaction.

Ben investigates the reaction between zinc and hydrochloric acid.

Hydrogen and zinc chloride are made in this reaction.

(a) Write down the word equation for the reaction.

_____ [1 mark]

(b) Ben uses the apparatus in the diagram. Look at the diagram.

Ben uses 0.5 g of zinc powder and 50 cm³ of hydrochloric acid. He tips the flask. The zinc falls into the acid. This starts the reaction. Every minute Ben measures the volume of hydrogen gas collected. He plots his results on a graph. Look at the graph.

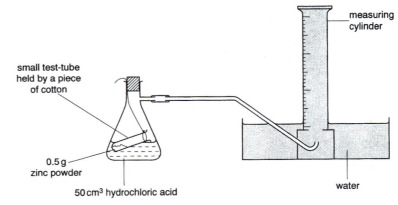

(i) Ben does the experiment a second time. He uses the same amounts of zinc and hydrochloric acid. This time he uses **warm** hydrochloric acid rather than **cold**. On the **grid**, sketch the graph of the results you would expect.

Label your line **T**. [2 marks]

(ii) Ben does the experiment a third time. He uses the same volume of hydrochloric acid but he only uses 0.25 g of zinc. When the zinc is all used up he only collects 92.5 cm³ of gas. Explain why.

_____

_____ [1 mark]

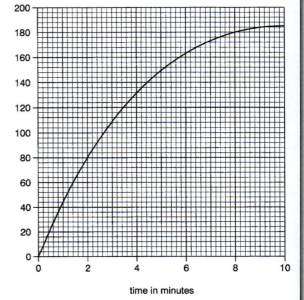

(c) The reaction between zinc and **dilute** hydrochloric acid is much slower than the reaction between zinc and **concentrated** hydrochloric acid.

Explain why using ideas about particles.

_____

_____ [2 marks]

(d) The reaction between zinc **lumps** and dilute hydrochloric acid is much slower than between zinc **powder** and dilute hydrochloric acid.

Explain why using ideas about particles.

_____

_____ [2 marks]

(e) Ben does some more experiments.

He wants to find a catalyst for the reaction between zinc and hydrochloric acid. He uses 0.5 g of zinc powder and 50 cm^3 of hydrochloric acid. He also adds 0.1 g of another substance. Ben measures the time it takes to collect 100 cm^3 of hydrogen. Look at his table of results.

Substance	Time to collect 100 cm^3 of hydrogen in seconds	Colour of substance at start of experiment	Colour of substance at end of experiment
No substance	150	–	–
Magnesium chloride	150	white	white
Copper (II) chloride	20	green	pink
Copper powder	15	pink	pink

Which **one** of the substances in the table is a catalyst? Explain your answer fully.

_____

_____

_____ [3 marks]

Answers are given on p.119

Exam Question and Answer

1) Use the periodic table on page 7 to help you answer these questions.

a) A Russian chemist named Mendeleef produced a periodic table. His periodic table had the elements in order of increasing atomic mass. Find the elements potassium and argon in the periodic table.

i) What problem is caused by using atomic mass to place these elements in order?

Argon has a smaller atomic mass so should be before potassium (1 mark) [1 mark]

ii) Show how this problem is solved for potassium and argon in a modern periodic table.

Elements are arranged in order of increasing atomic number (1 mark).
Potassium has an atomic number of 17 and argon of 18 (1 mark).   [2 marks]

The table below gives information about some elements in the third period of the periodic table.

b) There is a pattern in the electron arrangements of atoms of elements in this period.

i) Complete the missing electron arrangements in the table.   [2 marks]

ii) What is the connection between the electron arrangement and the position of the element in the periodic table?

Element	Symbol	Electron arrangement	Formulae of chlorides
Sodium	Na	2,8,1	NaCl
Magnesium	Mg	2,8,2	
Aluminium	Al	2,8,3	$AlCl_3$
Silicon	Si	2,8,4 (1 mark)	$SiCl_4$
Phosphorus	P	2,8,5	PCl_5
Sulphur	S	2,8,6	S_2Cl_2
Chlorine	Cl	2,8,7 (1 mark)	

The number of shells indicates which period (1 mark) and the number of outer electrons indicates which group the element is in (1 mark).   [2 marks]

c) There is a pattern in the formulae of chlorides in this period.
Suggest the formula for magnesium chloride. $MgCl_2$ (1 mark)   [1 mark]

d) Sodium reacts with cold water.

i) Write down the names of the products of this reaction.

Sodium hydroxide and hydrogen (1 mark)   [1 mark]

ii) Write a balanced equation for this reaction. $2Na + 2H_2O \rightarrow 2NaOH + H_2$ (2 marks)   [2 marks]

e) Potassium is in the same group of the periodic table as sodium.

i) Write down the electron arrangement in a potassium atom.   2,8,8,2 (1 mark)   [1 mark]

ii) Explain why potassium reacts faster than sodium with cold water.

Potassium atoms have a larger atomic radius than sodium atoms (1 mark)
so a potassium atom can lose an electron more easily (1 mark).   [2 marks]

How to score full marks

a) i) **A periodic table will normally have both atomic masses and atomic numbers included**. For most elements the order of atomic number is the same as that for atomic mass, but this is not true for potassium and argon.

ii) You need to refer specifically to potassium and argon. You could refer to **putting the elements underneath elements that have similar chemical properties**, e.g. potassium is placed under sodium in a modern periodic table because they both react rapidly with cold water.

b) i) **Only the number of outer electrons changes** from sodium to chlorine. For each element the number of electrons in the outer shell is the group number.

ii) The electron arrangement of an element will determine where it is found in the periodic table.

c) As you go **across period 3, one atom of the element combines with an extra chlorine atom**. You may also have spotted that the number of chlorine atoms combined matches the group number of the element.

d) i) Sodium is an alkali metal so it reacts with water to form an alkali, sodium hydroxide. **Don't write the formulae down**, the question asks for the names of the products.

ii) There is **no need to include state symbols** in this equation **because they are not asked for**. Any correct multiple of this equation could have been written. One mark is for writing the correct formulae of the reactants and the products, the other is for balancing. Remember not to change the formulae when you are balancing the equation.

e) i) You need to **know the electron arrangement for the first 20 elements** in the periodic table.

ii) **The question is worth 2 marks so you must make 2 distinct points**. One mark is for commenting on the loss of an electron from a potassium atom compared with a sodium atom. One mark is for explaining why this happens. You could refer to potassium atoms having more shielding shells of electrons.

Don't make these mistakes...

Elements are arranged in order of increasing atomic number in the periodic table and not atomic mass. You must know the difference between the atomic number and the atomic mass on a periodic table.

Only use the names chlorine, bromine and iodine when referring to elements. Chloride, bromide and iodide refer to the negative ions formed by these elements.

Atoms don't lose or gain an electron faster or slower than another. It is the ease of electron gain or loss (or amount of energy needed to add or remove an electron) that is important.

Don't use Cl to represent chlorine as an element; Cl refers to a chlorine atom. All of the elements in group 7 exist as diatomic molecules, so the formula for chlorine as an element is Cl_2.

Don't use the term inert gases for noble gases, because some group 0 elements can react.

Key points to remember

Elements are arranged in the periodic table so that it is easier to understand and learn the properties and reactions of the elements

Remember periods of elements go across the periodic table and groups of elements go down the periodic table. The group number of an element corresponds to the number of electrons in the outer shell of the electron arrangement. Elements have similar chemical properties when they have the same number of electrons in their outer shell.

The period number of an element corresponds to the number of occupied shells in the electron arrangement. Elements in the same period do not have similar chemical properties.

The number of electrons in an atom is the same as the atomic number of the element. If you choose the atomic mass you will get the electronic arrangement of an element wrong. You should know the electronic arrangement of the first 20 elements.

Make sure you know some uses of noble gases, halogens and their compounds, and alkali metals and their compounds.

Group 1 elements are called the alkali metals because they react with cold water to give an alkaline solution, the hydroxide, and hydrogen. Group 1 elements are very reactive metals that form ionic compounds. The reactivity of group 1 elements increases down the group as the atomic radius increases, so that loss of an electron from an atom becomes easier.

Group 0 elements are called the noble gases because they are very unreactive. Group 0 elements have a stable outer octet of electrons. They do not always have a full outer shell: the third electron shell can take up to 18 electrons so argon (2,8,8) does not have a full outer shell.

Group 7 elements are known as the halogens. They are reactive non-metals and form ionic compounds with metals and covalent compounds with non-metals. The reactivity of group 7 elements decreases down the group as the atomic radius increases, so the gain of an electron by an atom becomes more difficult.

You should learn the formulae of the following substances: sodium hydroxide, sodium chloride, potassium hydroxide, potassium chloride, chlorine, bromine and iodine.

The formulae of similar compounds are similar, e.g. potassium iodide is KI and lithium bromide is LiBr.

Transition elements are found in the centre of the periodic table. All transition elements are metals and have typical metallic properties. Compounds of transition metal compounds are often coloured.

Metals are found on the left-hand side and centre of the periodic table. Metals are lustrous, sonorous, ductile, malleable, have a high thermal and electrical conductivity, and normally have a high density, melting and boiling points. Metals form basic oxides.

Non-metals are found on the top right-hand side of the periodic table. Non-metals are not lustrous, not sonorous, not ductile, not malleable, are poor thermal and electrical conductors, and often have a low density. Non-metals form acidic oxides.

Chlorine, bromine and iodine are all in group 7 of the periodic table.

(a) Explain why the three elements are placed in group 7.

Use ideas about electronic arrangements.

_____ [1 mark]

(b) Chlorine is one of the elements in period 3 of the periodic table.

How can you tell from its electronic arrangement?

_____

_____ [2 marks]

(c) The chloride ion has the formula Cl^-.

   (i) How does a chlorine atom become a chloride ion?

_____ [1 mark]

   (ii) What is the formula for an iodide ion?

_____ [1 mark]

(d) Chlorine and bromine both react with iron.

Chlorine forms iron(III) chloride and bromine forms iron(III) bromide, $FeBr_3$.

   (i) Write the word equation for the reaction between chlorine and iron.

_____ [1 mark]

(ii) Write the balanced symbol equation for the reaction between iron and bromine.

_____ [2 marks]

(iii) Chlorine is more reactive towards iron than bromine.

Explain why. Use ideas about electron gain.

_____

_____ [1 mark]

(e) Chlorine reacts with aqueous potassium iodide solution to form iodine and aqueous potassium chloride.

Write the balanced symbol equation for this reaction.

_____ [2 marks]

(f) Astatine is below iodine in group 7.

Bromine is an orange liquid.

(i) Predict the physical appearance of astatine at room temperature.

_____

_____ [2 marks]

(ii) Predict how astatine will react with aqueous potassium iodide solution. Explain your answer.

_____

_____ [2 marks]

(g) Fluorine is the most reactive element in group 7.

Suggest why fluorine will not react with neon.

_____ [1 mark]

Answers are given on p.120

13 Chemical Calculations

Exam Question and Answer

1) Ammonia is made in the Haber process. Nitrogen, N_2, and hydrogen, H_2, react to make ammonia, NH_3. The reaction is reversible and forms an equilibrium mixture.

   Ammonia can be reacted with sulphuric acid to form a fertiliser. The name of the fertiliser is ammonium sulphate, $(NH_4)_2SO_4$.

a) Calculate the relative formula mass, M_r, of ammonium sulphate.

N 2 x 14 = 28
H 8 x 1 = 8
S 1 x 32 = 32
O 4 x 16 = 64
So M_r is 132 **(1 mark)**   [1 mark]

b) A third fertiliser is a compound containing only carbon, hydrogen, nitrogen and oxygen.
   Some of the fertiliser is analysed. It contains 24 g of carbon, 8 g of hydrogen, 56 g of nitrogen and 32 g of oxygen. There are 2 moles of atoms in 24 g of carbon.

   i) Calculate the number of moles of atoms in 8 g of hydrogen.

   8 ÷ 1 = 8 **(1 mark)**   [1 mark]

   ii) Calculate the number of moles of atoms in 56 g of nitrogen.

   56 ÷ 14 = 4 **(1 mark)**   [1 mark]

   iii) Calculate the number of moles of atoms in 32 g of oxygen.

   32 ÷ 16 = 2 **(1 mark)**   [1 mark]

   iv) Work out the formula of the fertiliser.

   CH_4N_2O **(1 mark)**   [1 mark]

c) A factory uses 14 tonnes of nitrogen every day. What is the maximum mass of ammonia that the factory can make in a day?

N_2 + $3H_2$ ⇌ $2NH_3$ **(1 mark)**
1 mole   3 moles   2 moles **(1 mark)**
28 g   6 g   34 g **(1 mark)**
So 14 tonnes   3 tonnes   17 tonnes **(1 mark)**   [4 marks]

a) You must be able to calculate relative formula masses if you are to tackle calculations in chemistry successfully. Make certain that you show **all your working out** and **line up the tens and unit columns** to avoid silly arithmetic errors.

b) This calculation is carefully structured to allow you to see the route through the question. Numbers in GCSE calculations often cancel easily, as here. If you **do not get simple whole numbers then you should check** to see if you have made a simple arithmetic error.

In (iv) the formula can be written in any atomic symbol order. Another acceptable formula is $C_2H_8N_4O_2$.

c) You will probably find calculations one of the hardest parts of the GCSE course. Make certain that you organise your answer carefully. Start with the balanced equation. The **amounts and masses are written directly underneath the correct formulae**. As in (b), the numbers cancel out easily. Do not worry if the question uses masses instead of grams, work out the question as though it is grams and then change your final answer to the required unit.

Don't make these mistakes...

You must show all your working out in calculations. Often there are marks for the working out. This means even if you get a calculation wrong you may get an error carried forward mark if you have only made one mistake. Examiners often give an error carried forward providing the working out is given and is correct.

Remember that you cannot have a percentage yield that is greater than 100%.

Don't use too few or too many significant figures in numerical questions. Normally three significant figures will do.

Always take a calculator into the exam so that you don't waste time doing the calculations long hand.

Check that your answer makes sense in case you have made a silly calculator error.

When using the periodic table to find a relative atomic mass, don't use the atomic number of an element by mistake.

Don't put down a unit for relative formula mass, it is just a number. The molar mass of a substance is the relative formula mass in grams.

When carrying out calculations with concentration of solutions, remember to use dm^3 as the unit of volume and not cm^3. $1000\,cm^3$ equals $1\,dm^3$.

Key points to remember

Percentage yield is a way of comparing the amount of the product made (actual yield) to the amount expected (predicted yield).

$$\text{percentage yield} = \frac{\text{actual yield}}{\text{predicted yield}} \times 100$$

The empirical formula is the simplest ratio of each different type of atom in a compound.

Compounds have a fixed composition by mass. This can be worked out from the relative formula mass.

In all calculations from equations there will be some marks for working out the relative formula masses of the reactants and products.

You don't have to remember relative atomic masses: they are always given in the exam paper.

The mole is a measure of the amount of a substance. One mole of a substance contains about 6×10^{23} particles.

You should remember the relationship between number of moles, volume of solution (in dm^3) and concentration using the triangle:

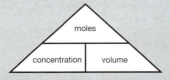

A solution is made up of a solute dissolved in a solvent. The concentration of a solute is measured in mols/dm^3: this shows how many moles of solute are in 1 dm^3 of solution.

One mole of a gas at room temperature and pressure has a volume of 24 000 cm^3. You do not have to remember this information: it will always be given in your exam paper. One mole of a pure substance has a mass of the relative formula mass in grams.

You should remember the relationship between the number of moles, relative formula mass and mass in grams using the triangle:

Fertilisers often contain the essential elements nitrogen, phosphorus and potassium

Practise:
- constructing balanced symbol equations
- working out the relative formula mass of compounds.

A balanced equation shows the ratio of moles of each reactant and each product.

Questions to try

Examiner's Hints

(b) Remember that one mole of a substance contains approximately 6×10^{23} molecules.

(d) Use the fraction of nitrogen in the total relative formula mass to work out the percentage of nitrogen.

Caffeine has the molecular formula $C_8H_{10}N_4O_2$.

(a) How many atoms are there in one molecule of caffeine?

_____ [1 mark]

(b) How many atoms are there in one mole of caffeine?

_____

_____ [2 marks]

(c) What is the relative formula mass of caffeine?

_____

_____ [2 marks]

(d) What is the percentage by mass of nitrogen in caffeine?

_____ [2 marks]

Examiner's Hints

● Remember that the symbols and atomic masses of gallium and phosphorus can be found in the periodic table.

● You should use the relationship between the number of moles, mass and the relative atomic mass.

The light emitting diode used in some calculator displays is made from a compound containing gallium and phosphorus.

A sample of this compound was analysed and found to contain 0.69 g of gallium and 0.31 g of phosphorus.

(a) How many moles of atoms are there in 0.69 g of gallium?

_____ [1 mark]

(b) How many moles of atoms are there in 0.31g of phosphorus?

_____ [1 mark]

(c) What is the formula of the compound?

_____ [1 mark]

Examiner's Hints

(b) (i) Remember to write down the equation, and underneath each substance write the appropriate number of moles and mass. You will need the relative formula masses for copper(II) oxide and copper(II) carbonate.

(b) (ii) You will need your answer to **(i)** and must choose the correct actual mass to use with the equation for percentage yield.

Gita makes some copper(II) oxide, CuO, from copper(II) carbonate, $CuCO_3$.
Carbon dioxide, CO_2, is the other product.

(a) Write down the symbol equation for this reaction.

_____ [1 mark]

(b) Gita uses 1.24 g of copper(II) carbonate and makes 0.72 g of copper(II) oxide.

   (i)   Gita predicts the mass of copper(II) oxide she should make.
        What is the predicted mass of copper(II) oxide?

_____

_____

_____ [3 marks]

   (ii) What is Gita's percentage yield of copper(II) oxide?

_____

_____ [2 marks]

   (iii) What volume of carbon dioxide measured at room temperature is made?

_____ [1 mark]

Answers are given on p.121

14 Electricity

Exam Question and Answer

1) Cuthbert has a car. He looks in his handbook. It shows a circuit diagram for the car's lights. The lamps are connected in parallel. Look at the circuit diagram for the car's lights.

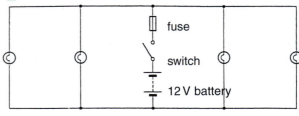

a) Explain why the lamps are connected in **parallel**.

When in parallel if one lamp fails the others remain lit. **(1 mark)** [1 mark]

b) There is a fuse in the circuit. It is used to protect the circuit. Look at the diagram. **Explain** how the fuse works.

If the current gets too high **(1 mark)** the fuse will melt **(1 mark)** and switch off the circuit **(1 mark)**. [3 marks]

c) The power of each working lamp is 24 W. Calculate the current flowing through each lamp.

power = voltage x current **(1 mark)**   current = 2 **(1 mark)**
current = power / voltage **(1 mark)**   A or amps **(1 mark)**
current = 24 / 12 **(1 mark)** [5 marks]

d) Cuthbert has some fuses: **2A 5A 10A 15A 20A 50A**
He uses the **10A** fuse in this circuit. Explain why this is a sensible choice.

There is an 8A (4 x 2A) current in the circuit **(1 mark)** The 10A fuse is just higher than that needed **(1 mark)** but not dangerously high (for the circuit) **(1 mark)**. [3 marks]

e) The car's electrical circuits are made of wire and they are connected to a battery. The battery contains a liquid which is an acid. Charge moves in the liquid and the wires. Explain how the charge moves.

Electrons carry charge in the wire **(1 mark)** and ions carry charge in the liquid **(1 mark)**. [2 marks]

f) The car stereo system is connected to the 12 V battery. It has a current of 4A when at full power. Calculate the resistance of the system.

resistance = voltage / current **(1 mark)**   resistance = 3 **(1 mark)**
resistance = 12 / 4 **(1 mark)** Ω or Ohms **(1 mark)** [4 marks]

g) Suggest which fuse Cuthbert could be used to protect the stereo system. Choose from **2A 5A 10A 15A 20A 50A**

5A fuse is best **(1 mark)** [1 mark]

How to score full marks

1) The context for this question is the electric circuits of a car. You should **not** be put off by this (if you know nothing about cars). Neither should you be **overconfident** (if you know lots about cars). The question is really about electricity and your understanding of it.

a) In most circuits lamps are connected in parallel so that **if one lamp fails the others remain on**. It may also be because **in parallel they can be controlled individually**. However, in this circuit there is only one switch, so that is not the case here. It is tempting to say that all lamps will get full brightness or voltage. This misses the point because the four bulbs here could be connected in series to a 48 V supply and they would all reach full brightness.

b) Look at the 'Don't make these mistakes' section in Chapter 17. This will give you more guidance on answering questions about fuses.

c) You need to remember and use several equations concerned with electricity. This one is to do with **power**. You can calculate the current flowing through each lamp by remembering this equation: **power = voltage × current**. It then needs to be rearranged to put current as the subject of the equation: **current = power/voltage**

d) When choosing which fuse Cuthbert should use, you need to know the current it needs to carry. **The best fuse is one that allows the current (8A) through**. A 2A and 5A fuse would melt and break the circuit. A 15A, 20A or 50A fuse is much too high and could be a fire hazard. With the 50A fuse, for instance, it would be very dangerous for a possible 49A current to flow through the radio. This could cause a fire. **A 10A fuse would only allow 2A more than normal before it melted and broke the circuit**. This protects the circuit better than the bigger fuses.

e) Electric current is a flow of charge. The charges are carried differently in solids and liquids. In **liquids they are carried by ions**. In **solids they are carried by electrons**.

Don't make these mistakes...

If you can't calculate the current in (c), don't worry. You may still gain the marks in (d) as long as you state a reasonable current and explain it correctly.

Often you can gain marks easily by writing down the correct units. With electricity many candidates are not too sure of their units. Some examiners know this and may focus on them. Try to be an expert on electrical units by learning them and practising them. They are probably not one of the most important aspects of science but they will get you easy marks if you know them.

Don't use vague words like 'electricity' in your explanations. Use words such as 'current' correctly instead. Try not to get confused about the way a fuse works. A fuse does not work like a battery or circuit breaker (see Chapter 17).

Key points to remember

Units

pd or voltage (volts, V)

resistance (ohms, Ω)

current (amps, A)

charge (coulombs, C)

power (Watts, W)

time (seconds, s)

energy (joules, j)

When electrons are removed or added from an insulator it becomes charged (static electricity). This is usually caused by rubbing two insulators together, for example a balloon and an acrylic jumper (the balloon is then attracted to some objects). Some times these charges will build up and a current flows in the form of a spark. This can be dangerous when there are flammable materials about. Static electricity can be useful, for example in photocopiers, precipitators (these remove dust and particles of dust and are used in chimneys) and spraying paint or crops.

When spraying a car the paint is charged positive. These charges in the droplets repel each other and the drops break up making the spray finer (this means less paint is needed). The car is charged negative and the positive drops of paint are attracted to the car. This helps to attract the paint to parts of a car that cannot be easily reached (this makes sure that all parts of the car are painted).

New York at night

When a voltage is applied to devices they do not always behave the same.

Current electricity can be provided by cells or batteries (d.c. or direct current)

Equations

power (W) = current (A) $\times$ voltage (V)

resistance (W) = voltage (V) / current (A)

charge (C) = current (A) $\times$ time (s)

power (W) = energy (j) / time (s)

The power (W) tells you how much energy an appliance transfers each second. A 100 W bulb transfers 100 j of energy every second. There are 1000 W in 1 kW. Large powered appliances can use a lot of energy each second and can be costly to run.

There are two types of charge: positive and negative. Positive–negative charges attract each other. Negative–negative or positive–positive charges repel each other

a.c., or alternating current, is provided by many power supplies (which contain transformers). The mains supply in your home is a.c.

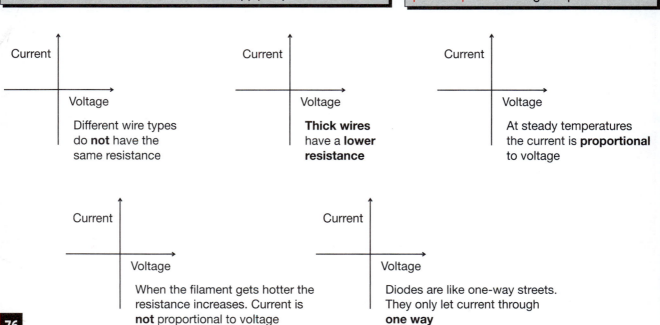

Current

Voltage

Different wire types do **not** have the same resistance

Current

Voltage

Thick wires have a **lower resistance**

Current

Voltage

At steady temperatures the current is **proportional** to voltage

Current

Voltage

When the filament gets hotter the resistance increases. Current is **not** proportional to voltage

Current

Voltage

Diodes are like one-way streets. They only let current through **one way**

Examiner's Hints
● You need to know about how **insulators become charged**, the **risks** and **uses**.
● You need to be able to do some electrical calculations. Make sure **you know your formulae and units** before you start.

---

(1) Chris does an investigation with electrical circuits. Look at the diagram.

○ 12V d.c. ○

A

sliding connector

0 cm   50 cm   100 cm

resistance wire

(a) He slides the connector towards 100 cm. The bulb dims. Explain why.

As the resistance increases the current
decreases. [2 marks]

(b) He replaces the resistance wire for a thinner piece. It gets hot but does not melt.

Look at the graphs of the two results. The thinner wire's results fall on a **curve**. Explain why.

current

Thick
Thin

0   length   100

_____
_____ [3 marks]

(c) The maximum current through the lamp is **2A** and the voltage is **12V**.

 (i) Calculate the **power** of the lamp. Answer ___24___ Units ___Watts___
[4 marks]

 (ii) Calculate the **resistance** of the lamp. Answer ___6___ Units ___Ω___
[4 marks]

(2) (a) Static electricity can be useful or dangerous.

 (i) It can be **useful** when spraying cars with paint. Explain why it is useful **and** how it works.

_____
_____
_____
_____ [6 marks]

(ii) Static electricity can be **dangerous** when refuelling an aircraft. Explain how.

_____

_____ [3 marks]

(b) Eli rubs a plastic ruler with a cloth. She holds the ruler over some small pieces of paper. Look at the diagram.

paper

plastic strip

What happens to the paper? Explain why.

The paper rises to the strip because the strip is charged

_____ [3 marks]

(c) Eli has a nylon carpet in her room. She wears acrylic slippers and walks on the carpet. She gets a shock when she touches the metal radiator. Explain why.

earthed metal radiator

nylon carpet

acrylic slipper

_____

_____

_____ [4 marks]

Answers are given on p.122

Exam Question and Answer

1) a) The diagram shows a simple transformer.

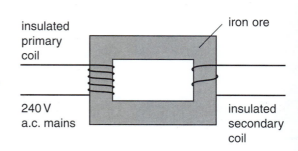

insulated primary coil

iron ore

The primary coil is attached to a 240 V a.c. supply.

240 V a.c. mains

insulated secondary coil

   i) A voltage is produced in the secondary coil.

     Explain how.

> The voltage produces a current in the primary coil causing a magnetic field **(1 mark)** which changes because of the a.c. **(1 mark)**. This changing field cuts the coils on the secondary inducing a voltage **(1 mark)**. [3 marks]

   ii) John wants to light a 12 V bulb using the 240 V mains. He uses a transformer connected to the 240 V mains supply. The primary coil has 4000 turns. Calculate the number of turns on the secondary coil.

> $V_p = 240V$ $V_s = 12\ V$ $N_p = 4000$ $N_s = ?$ **(1 mark)**
>
> The ratio of voltages = the ratio of the turns **(1 mark)**
>
> $\frac{V_s}{V_p} = \frac{N_s}{N_p}$ so $N_s = \frac{V_s}{V_p} \times N_p = 12/240 \times 4000 = 200$ turns **(1 mark)**

[3 marks]

2) a) A wind turbine can be used to make electricity.

a.c. generator

blades

The blades are joined to a simple a.c. generator.

The wind makes the blades go round.

This makes electricity.

Look at the diagram.

Explain how the generator makes the electricity.

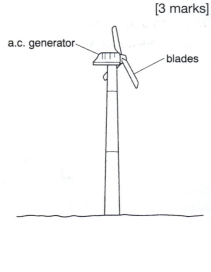

> Inside the generator there is a magnet **(1 mark)** and a coil **(1 mark)**. The coil spins **(1 mark)** inside the magnetic field **(1 mark)** and electricity is generated. [4 marks]

  b) Mains elecstricity is a.c.

   i) What does a.c. mean? <u>a.c. means alternating current **(1 mark)**.</u> [1 mark]

   ii) What kind of electricity do you get from a battery? <u>d.c. or direct current **(1 mark)**.</u> [1 mark]

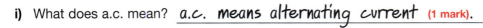

c) Coal and gas are fuels. They are used to make electricity in power stations.

    i) What happens to the fuel in a power station? <u>The fuel is burned</u> **(1 mark)**   **[1 mark]**

Power stations are connected to the National Grid.

    ii) What else is connected to the National Grid? Homes and factories **(1 mark).**   **[1 mark]**

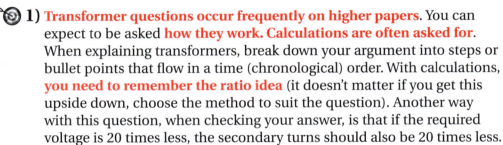

How to score full marks

1) **Transformer questions occur frequently on higher papers.** You can expect to be asked **how they work. Calculations are often asked for.** When explaining transformers, break down your argument into steps or bullet points that flow in a time (chronological) order. With calculations, **you need to remember the ratio idea** (it doesn't matter if you get this upside down, choose the method to suit the question). Another way with this question, when checking your answer, is that if the required voltage is 20 times less, the secondary turns should also be 20 times less.

2 a) **The question tells you that there is a generator in the wind turbine.** Although you cannot see the generator you need to **describe one or two of its parts and explain what they do.** Maybe think back to when you made one in class. A concise answer such as 'the coil spins in a magnetic field' will gain all 4 marks.

    b) You need to remember **a.c. = alternating current** and **d.c. = direct current**

    c) You will **often come across questions about the National Grid and power generation.** Be sure to know the key ideas in this area.

Don't make these mistakes...

You must show your working when doing transformer calculations. You may be throwing away valuable marks if you don't.

Avoid restating the question. For example, when describing the generator in the wind turbine, think of the main parts of the generator and describe what they do. Don't write vague comments like 'the wind spins the turbine which makes electricity in the generator'. This will score no marks at all because there is no science in the answer.

a.c. = alternating current. Write it clearly and avoid errors like 'alternative current'.

Lots of things are connected to the National Grid. Think big and avoid writing down the name of an appliance, such as a TV. Writing down homes, villages, towns or factories will convince the examiner you know what you're talking about.

Power supplied = voltage × current

Power loss = current^2 × resistance

So if the current increases by 10 times, the power loss is 100 times greater. It is sensible to keep the current as low as possible.

A motor has a coil that carries current, making it magnetic. This is affected by the field of the permanent magnet, and the forces make it spin. The motor will have more force if there are stronger magnets, more turns on the coil, more current or a soft iron core in the coil. The motor effect is also responsible for how loudspeakers work.

In generators the coil spins in a magnetic field. The generator will produce more current if there are stronger magnets, more turns on the coil, a bigger area in the coil or quicker movement.

These things can become magnetic: iron, steel, nickel and even a wire carrying a current. They can be forced to move when they are near another magnetic field.

Magnets have two poles, North and South. Opposites attract and like poles repel.

An electromagnet is a coil of wire around a soft iron core. Soft iron is ideal for an electromagnet because it easily looses and gains its magnetism when switched off or on. Learn the simple applications of an electromagnet in a scrap yard, circuit breaker, relay and electric bell.

Transformers have a primary (input) and a secondary (output) coil. If there are more turns on the secondary coil it is a step-up transformer and the voltage output is increased. They need a changing magnetic field to work and so they only work with a.c. current.

Transformers can get hot and make a buzzing (vibration) noise. Because this noise and heat require energy, they leave less energy through electricity in the secondary coil. The core is laminated so that eddy currents are reduced (but not eliminated). Sometimes a few more turns are required on the secondary coil than you might expect, to allow for the energy wastage.

The National Grid is a network of power stations, pylon cables, transformers, towns, homes and industry. Transformers are used to step-up or step-down the voltages. When transmitted at high voltage the current is small. This means that thinner cables can be used and there is less energy transferred from the cables as heat. This also reduces costs.

Power can be generated by power stations anywhere and then connected to the grid. a.c. is used because the voltages can be easily changed by transformers (remember that transformers do not work with d.c.). High voltages are used for transmission but they are then stepped-down so that they can be safely used in homes and factories.

Remember that in a transformer, the ratio of voltages = the ratio of turns

$$\frac{V_s}{V_p} = \frac{N_s}{N_p} \text{ or } \frac{V_p}{V_s} = \frac{N_p}{N_s}$$

For motors, the input is electricity and the output is movement. The opposite is true for generators, the input is movement and the output is electricity.

The motor effect is how loudspeakers work

(1) (a) Electric motors produce movement from electricity.

   (i) Explain how a motor works.

_____

_____

_____ [4 marks]

   (ii) Explain what you could do to make the motor spin faster.

_____

_____ [3 marks]

(b) A generator produces electricity from movement.

   (i) Explain how a generator works.

_____

_____

_____ [4 marks]

   (ii) Explain what you could do to make it produce more current.

_____

_____ [3 marks]

(c) Transformers are used in the National Grid.

   (i) Transformers are of **two** types: step-up transformers and step-down transformers. Explain the **differences** between step-up transformers and step-down transformers.

_____

_____ [2 marks]

(ii) Because of energy efficiency, sometimes there are, for a given voltage, more turns on the secondary coil than you would expect. Explain why.

_____

_____

_____

_____ [3 marks]

(2) (a) Write down, in words, the equation connecting the number of turns in each coil of a transformer and the voltages across them.

_____ [1 mark]

(b)

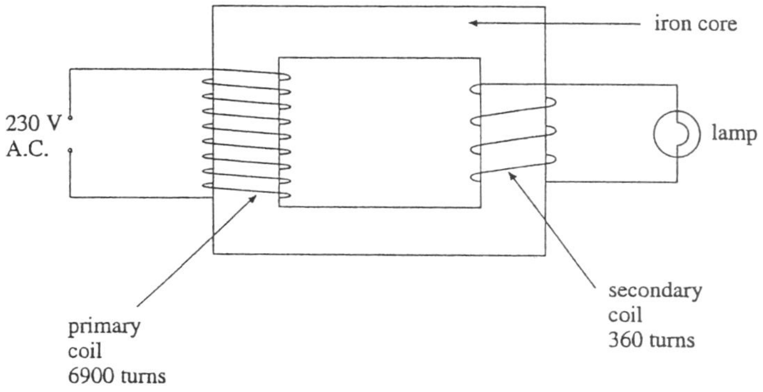

iron core

230 V A.C.

lamp

primary coil 6900 turns

secondary coil 360 turns

Use the information given in the diagram to calculate the voltage across the lamp.

_____

_____

_____ [3 marks]

(c) Explain why the transformer only works on alternating current (a.c.)

_____

_____ [2 marks]

(d) At the power station, voltages are increased to very high values using a step-up transformer. Explain why.

_____

_____ [2 marks]

Answers are given on p.123

Exam Question and Answer

1) A car driver stops a car by pushing down on the brake pedal. The brake pedal is connected to a lever which pushes the master piston. The master piston pushes a liquid through tubes to operate the brakes.

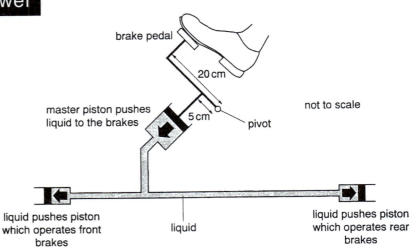

brake pedal

20 cm

master piston pushes liquid to the brakes

5 cm

pivot

not to scale

liquid pushes piston which operates front brakes

liquid

liquid pushes piston which operates rear brakes

a) **i)** The brakes will not work if there are bubbles of air inside the liquid. Explain why.

Air can be squashed and pressure cannot be passed through it easily (1 mark).   [1 mark]

ii) What does this tell you about the particles in a liquid compared to the particles in a gas?

Particles in a gas are further apart (1 mark).   [1 mark]

b) The driver's foot presses on a lever to make the brakes work. The pedal is connected 20 cm from the pivot. The master piston is connected 5 cm from the pivot. How big is the force on the master piston compared to the force of the driver's foot?

The force is bigger on the master piston (1 mark). It is 4 times bigger because the distance from the pivot is 4 times shorter (1 mark).   [2 marks]

c) The force on the piston is 1080 N. The piston has an area of 12 cm^2. Calculate the pressure on the liquid.

Pressure = Force/area (1 mark)   Pressure = 1080/12 (1 mark)

Pressure = 90 N/cm^2 (1 mark)   [3 marks]

d) To stop the car, the brakes have to transfer the kinetic (movement) energy of the car. What happens to this energy?

The energy heats the brakes by friction (1 mark).   [1 mark]

e) A car driver puts on the brakes. The car travels 40 m before it stops. The total braking force is 2000 N. Calculate the work done (energy transferred) by the brakes.

Work done = force x distance (1 mark)   Work done = 2000 x 40 (1 mark)

Work done = 80 000 j (1 mark)   [3 marks]

f) The front of the car is designed as a crumple zone. How does this help in a crash?

The kinetic or movement **(1 mark)** _energy is absorbed_ **(1 mark)**_, the crash time increases_ **(1 mark)**_, the acceleration is reduced_ **(1 mark)** _and the force is smaller_ **(1 mark)**.

[5 marks]

How to score full marks

a) ii) The question asks for **a comparison** so you must make this clear. Particles in a gas are 'far apart' is not enough.

b) Numbers are given in the question so you can use them to work out your answer. The fact that the question is worth 2 marks means you must **include 2 ideas for a full answer.**

c) You need to give **both** the correct answer (90) **and** the correct unit (N/cm^2) to score the third mark. When doing calculations, **always show your working**. If you make a mistake you can usually gain 1 or 2 marks even if the final calculation is wrong. If you show your working you are also less likely to go wrong.

e) Be sure to **use the correct formula** in the correct circumstances.

f) There are 5 marks here so the **examiner is looking for 5 good scientific ideas**. The idea of energy absorption is the easiest marking point to get. The other ideas are more advanced and need a higher level of scientific knowledge.

Don't make these mistakes...

Don't get confused between air and space. There **is not air** between particles in a gas. There **is space** (or nothing) between them.

Don't make silly mistakes when writing formulae:
- learn and remember them
- learn and practise when to use them
- learn and practise how to use them.

Don't forget your units. Always remember to learn your units, such as pressure in N/cm^2.

Don't forget to write down all your working for calculations. This will help you think clearer and will often increase your marks.

Key points to remember

Essential formulae

Force (N) = mass (kg) × acceleration (m/s^2)

Work done (j) = force (N) × distance (m)

Kinetic energy (j) = $\frac{1}{2}$ × mass × velocity^2 = $\frac{1}{2}$ × m × v × v

Momentum (kg m/s) = m × v

Speed (m/s) = distance/time

Acceleration (m/s^2) = change in speed/time

Distance travelled = area under a velocity time graph

Pressure (N/cm^2) = force/area

$P_1 × V_1 = P_2 × V_2$

Momentum before = momentum after

Speed is how far something travels in a second.

Thinking distance is the distance travelled when the driver is reacting. It can be affected by alcohol, drugs, poor concentration and the speed the car is moving.

Stopping distance = thinking distance + braking distance

Velocity is speed in a straight line.

Information about motion can be easily and clearly displayed in graphs. Practise how to draw and interpret them. Be sure to spot the difference between distance–time graphs and velocity–time graphs.

In car crashes these things reduce injuries: seat belts, crumple zones, air bags. They all increase stopping time but decrease the acceleration and forces needed.

When cars move they have (kinetic) energy. This energy needs to be transferred when stopping. It is transferred by friction in the brake pads through heat. The car brakes get very hot when used. In a crash the energy is also transferred through sound, heat and mainly through changing the shape of the car! Crumple zones absorb this energy by changing shape. The safety cage in a car (passenger compartment) is made of hard steel. The safety cage does not easily change shape in a crash.

Car brakes use hydraulics. Brake pipes are filled with a liquid oil. The oil cannot be compressed. It transmits pressure from the master cylinder (small area piston behind the foot brake) to the brake cylinder (large area piston near the brake disc). This increase in area makes the force much bigger but the pressure remains the same in the fluid.

Acceleration is how much the velocity (or speed) changes in a second.

Air bags inflate with gas in a crash. Gases can change shape easily because the particles can be pushed closer together. They absorb the energy of the moving driver in a crash and reduce injuries.

Braking distance is the distance moved when the brakes are on. It can be affected by speed, poor tread on tyres, wet, icy or slippery roads.

The graph shows the motion of a lorry.

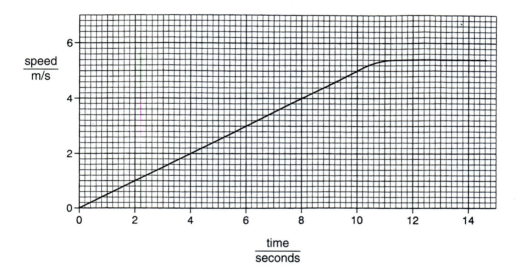

(a) Calculate the distance travelled by the lorry in the first 10 seconds.

_____

_____

Answer _____ m _____   [2 marks]

(b) Calculate the average speed of the lorry during the first 10 seconds.

_____

_____

_____

_____

Answer _____ Units _____   [3 marks]

(c) Use the graph to calculate the acceleration of the lorry during the first 10 seconds.

_____

_____

_____

_____

Answer _____ Units _____   [3 marks]

(d) The force accelerating the lorry is 3500 N.

Calculate the mass of the lorry.

_____

_____

_____ [3 marks]

Answers are given on p.124

Exam Question and Answer

1) Dave has an electric heater. The heater is very hot. Look at the diagram.

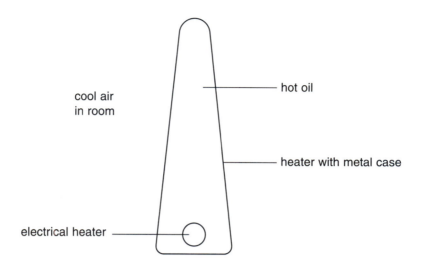

cool air
in room

hot oil

heater with metal case

electrical heater

a) Energy is transferred to the air in the room. Explain how. Use ideas about particles in your answer.

The hot oil particles in the heater move around quickly **(1 mark)**. They collide with the metal particles and make them vibrate more **(1 mark)**. These vibrations pass along the metal particles **(1 mark)**. The outside metal particles vibrate more **(1 mark)**. Cold air particles hit the hot metal particles and bounce off with more kinetic energy than before **(1 mark)**.   **[5 marks]**

b) The heater has a metal case. It has three wires in its cable and it has a 13A fuse. The heater has a fault. Explain how the wire fuse works.

When too much current **(1 mark)** goes through the fuse it heats up and melts **(1 mark)**. This breaks the circuit and cuts off the supply **(1 mark)**.   **[3 marks]**

c) Dave's hairdryer is double insulated. It has two wires in its cable.

   i) Write down the name of the two wires.

   The two wires in a double insulated appliance are live and neutral **(1 mark)**.
                                                                        [1 mark]

   ii) Double insulated appliances do not need earthing. Explain why.

   Double insulated appliances have an insulating case **(1 mark)**. This case cannot become live **(1 mark)**.   **[2 marks]**

How to score full marks

a) In order to **gain the full 5 marks, you need to make 5 separate points** about the behaviour of the particles. The question tells you to **use the idea of particles in your explanation**. You need to describe how the particles transfer the energy by passing on their kinetic energy (movement) to each other. In solids the particles are held together so they can only vibrate around fixed positions.

b) The question asks you **to explain** and in order to gain **the full 3 marks you must make 3 clear separate points**. Don't over-simplify your answers. It is the only chance you have to show what you know.

c) i) The examiner wants you to **name** the two wires as **live** and **neutral**. Both answers will be needed for 1 mark. Avoid putting the colours of the wires, brown and blue, because that is not answering the question and you will not score any marks even though it is correct.

ii) You are, again, being asked **to explain**. As the question is worth 2 marks, **it is best to write two sentences, each making a clear point**.

Don't make these mistakes...

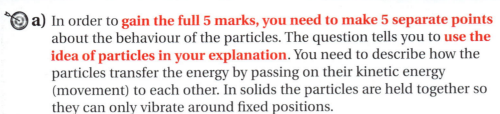

Don't use vague words like 'electricity' in your explanations. Use words correctly, such as 'current', instead. Also be sure to stress the comparison and say when too much current passes through.

Avoid saying that 'fuses blow up' or 'the wire snaps'. Do write down that 'fuses blow' or 'the wire melts'.

Don't confuse a fuse with a battery.

Particles cannot move 'around' in solids. They are in fixed positions but they can vibrate a little when cool and a lot when hot.

Heat particles do not exist!

Particles cannot 'vibrate' in fluids (liquids or gases). They are not in fixed positions but they can move around. They move around slower when cool (less kinetic energy) and faster when hot (more kinetic energy).

Don't confuse a fuse with a circuit breaker and say that the fuse 'trips'.

Don't write that particles are still when cold. This is incorrect.

Don't say that a fuse 'controls the current' in a circuit. A fuse cuts off the circuit when too much current passes because the wire melts.

When explaining energy transfer using the idea of particles, you must mention them and avoid writing vaguely about conduction and convection. Questions like these asking for particulate answers are at a higher level and require more advanced answers. Good low level answers may not gain any marks.

Key points to remember

Safety
- The live wire is brown, the neutral wire is blue, and the earth wire is green and yellow.
- Double insulated appliances do not need earthing.
- When too much current flows the fuse blows. The fuse wire gets hot, melts and cuts off the circuit. This can protect the appliance and can protect people.
- Double insulated appliances have two wires, live and neutral.
- Double insulated appliances have a plastic insulating case that cannot become live.

Insulators
- Good electrical conductors (or bad insulators) = copper, aluminium, gold.
- Good electrical insulators (or bad conductors) = glass, plastic.

Radiation (infra red) travels at the speed of light and is emitted easily from black dull hot objects but not from shiny silvered ones. It is absorbed easily by black dull hot objects but not by shiny silvered ones.

Evaporation happens when fast (hot) particles escape the liquid leaving the cooler particles behind. The hot particles get their extra kinetic energy from the surroundings which cool.

The Sun is the source of all energy on Earth

Three effects of electricity
- Magnetic (electromagnetism and electric motors).
- Heating (space heaters, immersion heaters, the wire in a light bulb gets so hot that it glows white hot).
- Chemical (charging a battery, electroplating a metal).

Conduction
- Good thermal conductors (or bad insulators) = copper, aluminium, gold.
- Good thermal insulators (or bad conductors) = glass, plastic, wood, polystyrene foam.

Particles
- Particles move more (or try to move more) when hot.
- The temperature is the average kinetic energy (KE) of the particles.
- Particles move around more in gases and liquids when hot.
- Particles vibrate more (in fixed positions) in solids when hot.
- Particles move with more kinetic energy when hot (they move around more in liquids and gases but vibrate more in solids).
- Particles cannot move through solid materials.
- Heat particles do not exist!
- For conduction in solids, particles vibrate more and the vibrations are passed through.
- For convection, fluids get warm, become less dense (particles get more spaced out) and rise.

Costs
High wattage = high cost to run
kWh units = power (Watts) × time (hours)
Cost (p) = price per unit (p) × units used (kWh)

Formulae
pd (volts) = current (amps) × resistance (Ohms)
power (Watts) = pd (volts) × current (amps)

Questions to try

Examiner's Hint

● The question invites you to use the idea of particles in your answer. This is essential. With evaporation, you need to write about how the particles behave when they evaporate or when they are left behind.

This question is about keeping things cool on a hot sunny day.

(a) Dennis has a drink cooler.

It has a porous pot.

It keeps his drink cool.

Look at the diagram.

water

drink

porous pot

The water soaks through the porous pot and evaporates. Explain how this keeps the drink cool. Use the idea of particles in your answer.

_____

_____

_____ [3 marks]

(b) Dennis and Dave make some sandwiches. They wrap them up in different ways. Look at the diagrams.

Dennis's sandwiches

black polythene bag

Dave's sandwiches

shiny aluminium foil

They leave their sandwiches in the sun. Dave's sandwiches stay cooler. Explain why.

_____

_____

_____ [3 marks]

Questions to try

Examiner's Hint
● Remember that conduction takes place in solids and convection in liquids and gases.

A saucepan is made of steel. It has a wooden handle. Look at the diagram.

— wooden handle
— water
— potatoes
— steel pan
— electric hot-plate

(a) Explain how energy is transferred from the electric hot-plate to the potatoes inside the pan. Use the idea of particles in your answer.

_____

_____

_____ [3 marks]

(b) Using a lid on the saucepan can reduce the energy needed to cook the potatoes. Explain why.

_____

_____

_____ [3 marks]

(c) The electric hot-plate has an average power rating of 2 kW for 20 minutes. The cost of electricity is 10p for 1 kWh.

   (i) Calculate the energy transferred by the hot-plate in 20 minutes.

   _____

   _____

   _____

   Answer _____ Units _____ [4 marks]

   (ii) Calculate the cost of the energy transferred in 20 minutes.

   _____

   _____ [2 marks]

Answers are given on p.125

Exam Question and Answer

1) Doctors use waves to help patients in hospitals.

a) Doctors use X-rays to look for broken bones. Doctors do **not** use gamma rays to look for broken bones. Explain why.

Gamma rays are not absorbed by bones **(1 mark)**.   [1 mark]

b) Doctors use ultrasound to check unborn babies. Doctors do **not** use X-rays to check unborn babies. Explain why.

X-rays can damage cells **(1 mark)**.   [1 mark]

c) Doctors use tracers to look for cancers in the body. They use a radioactive material that emits gamma rays. Explain how.

The radioactive material is injected/swallowed **(1 mark)**. This is passed round the body **(1 mark)**. The radioactive material is absorbed by the cancer **(1 mark)**. More gamma radiation is emitted from the cancer and this is detected outside the body **(1 mark)**.   [4 marks]

d) Explain why gamma radiation is used to sterilise surgical instruments.

Gamma radiation kills living cells/microbes **(1 mark)**.   [1 mark]

2) A ship uses an ultrasound transmitter to study the sea bed. Look at the diagram.

a) What is ultrasound?

It is sound that is very high frequency, too high pitched for humans to hear **(1 mark)**.   [1 mark]

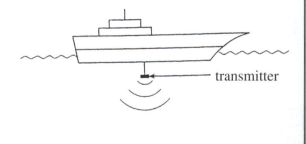

transmitter

b) The speed of the ultrasound waves is 1500 m/s. The wavelength of the ultrasound waves is 0.001 m. The waves take 2 s to travel to the sea bed, reflect and return to the ship.

   i) Calculate the frequency of the ultrasound waves.

frequency = wave speed/wavelength **(1 mark)** frequency = 1500/0.001 **(1 mark)**
frequency = 1 500 000 **(1 mark)** hertz **(1 mark)**   [4 marks]

   ii) Calculate the depth in metres of the sea below the transmitter.

distance = speed x time **(1 mark)**   distance = 1500 x 2 **(1 mark)**
distance = 3000 **(1 mark)** depth of sea = 1/2 this distance
= 1500 m **(1 mark)**   [4 marks]

How to score full marks

1) This question is testing your knowledge of the **electromagnetic spectrum. You must know about the order and nature of the waves.**

a) X-rays are used to look for broken bones because **they penetrate flesh but are absorbed by bone.** You need to write about gamma rays making the point that they are too penetrative and pass easily through flesh and bone. This is because gamma rays usually have a higher frequency and shorter wavelength than X-rays.

b) Ultrasound waves are **not** part of the electromagnetic spectrum. **X-rays** are electromagnetic and **will damage and kill living cells**. They can be very harmful to unborn babies. You could also have written that X-rays cause cancer.

d) The question asks **why gamma** is used. When talking about sterilisation by gamma radiation, **focus on what the radiation does**. If you said that it 'cleaned' the instruments that would be too vague. Give **a scientific answer** and say it **kills the cells or microbes**.

2) a) Ultrasound is **very high frequency sound** above the range of human hearing (about 20 000 Hertz). When used to check unborn babies, it passes into the woman and reflects from the different layers inside. The sound is detected on its return and a picture of the inside of the woman can be produced.

b) i) Be sure of the wave equation, **velocity = frequency × wavelength**. This is an easy way to gain 4 marks. Remember you need to **show all your working**. This can help you think more clearly, get you to the right answer and if you go wrong you will gain marks for some of your efforts. Remember that **frequency is measured in Hertz (Hz)**. This is one of the units that can be easily forgotten in an exam.

ii) **Distance = speed × time**. This equation can be used in lots of contexts: Force and Motion, The Earth and Beyond and here in Waves to do with the speed of sound. Again, as in all calculations, **remember to show all your working**.

Don't make these mistakes ...

Don't forget the equation Speed = Distance/Time. You will sometimes need to rearrange the equation if you need to know what distance or time is. You need to know how to rearrange equations, or remember them. Be sure to rearrange them correctly.

Don't forget to use the 'triangle' method if you need to.

Distance = Speed × Time
Time = Distance /Speed

Avoid saying things like 'X-rays can see bones'. This is not scientific enough.

You need to remember the wave equation and how to use it.

wave Speed = Frequency × Wavelength
Frequency = wave Speed/Wavelength
Wavelength = wave Speed/Frequency

When medical tracers are used they give out gamma radiation, which leaves the body where it is detected by a camera. Don't confuse the tracer idea with radiotherapy. Radiotherapy is when gamma radiation is 'fired' inside the body to hit a cancer and kill its cells.

Key points to remember

Waves

Be sure to learn that sound is a longitudinal wave that travels slower than light and, like all waves, can be reflected, refracted and diffracted.

Recognise transverse waves, learn these features and understand what they mean:

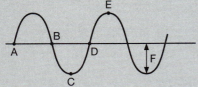

wavelength = A – D

crest = E

trough = C

amplitude = F

Recognise longitudinal waves and know about rarefactions and compressions.

Know that the frequency of ultrasound is too high for humans to hear.

Know how ultrasound is used in medicine to scan unborn babies or break-up kidney stones.

Know about p and s seismic waves travelling through the Earth and how they provide evidence of the Earth's structure.

Know that waves carry energy and not matter.

Know what happens to the frequency of waves when their wavelength increases or decreases.

Know and use the wave equation:
Wave speed = frequency × wavelength s = f × W

Electromagnetic spectrum

Know about the parts and order of the electromagnetic spectrum.

long wavelength						short wavelength
radio	microwaves	infra-red	visible	ultraviolet light	X-rays	gamma rays
low frequency						high frequency

Know some of the uses and dangers of these waves. For example:
- X-rays can be used to look inside the human body but can destroy living cells
- ultraviolet causes sunburn and sometimes skin cancer
- gamma rays destroy living cells and so can be used to treat cancer.

Learn and understand the use of tracers in medicine.

Light

You can see objects because they reflect light into our eyes and you need to be able to draw ray diagrams of light rays being reflected by plane and curved mirrors (including the normal line). Be sure, also, to know what plane mirrors and curved mirrors are used for. (Convex: car door mirror, security mirror. Concave: dentists magnifying mirror.)

Practice drawing diagrams of light rays being refracted and know what happens to light rays refracted by a lens. Converging lenses focus light by refraction.

Know how light travels along optical fibres by internal reflection and how it can be used as an endoscope by doctors or as part of a cable communication system.

Understand how lenses work in cameras and the human eye.

Questions to try

Q1–3

Examiner's Hints
- You need to know about the electromagnetic spectrum, the two types of seismic waves, and about ultrasound waves. Know and use the wave equation:
  wave speed = frequency × wavelength   s = f × W

(1) (a) Here is a list of waves.

gamma   infra-red   visible light   microwaves   radio   sound   ultra-violet   X-rays

   (i)   Which wave has the shortest wavelength? _____ [1 mark]

   (ii) Which wave has the highest frequency? _____ [1 mark]

   (iii) Which wave is **not** part of the electromagnetic spectrum? _____ [1 mark]

(b) What do all parts of the electromagnetic spectrum have in common?

_____ [1 mark]

(c) Microwaves are used to heat food. Explain how they raise the temperature of food.

_____

_____ [2 marks]

(d) High levels of gamma rays are used to treat cancer. Look at the diagram. Gamma rays kill cancer cells. Unfortunately they also kill healthy cells. Doctors reduce this problem in the following way.

area of cancer

gamma rays

patient

gamma ray source

the source moves around the patient in a circle

The gamma ray source rotates in a circle. The cancer is at the centre of this circle. This kills cancer cells but reduces damage to healthy cells. Explain how.

_____

_____

_____ [2 marks]

(2) Doctors use ultrasound in hospitals.

(a) Ultrasound waves are used to scan unborn babies. Explain how.

_____

_____

_____ [4 marks]

(b) It is an advantage to use ultrasound waves to cure problems
from kidney stones. Explain how they are used and why using them
is an advantage.

_____

_____ [3 marks]

(c) An ultrasound wave has a frequency of 30 000 Hz and travels at a speed of 1 500 m/s.

(i) Calculate the **wavelength** of the wave.

_____

_____

_____

Answer _____ Units _____ [5 marks]

(ii) The doctors cannot hear the ultrasound wave. Explain why.

_____

_____ [2 marks]

(3) Seismic waves pass through the Earth.
Look at the diagram.

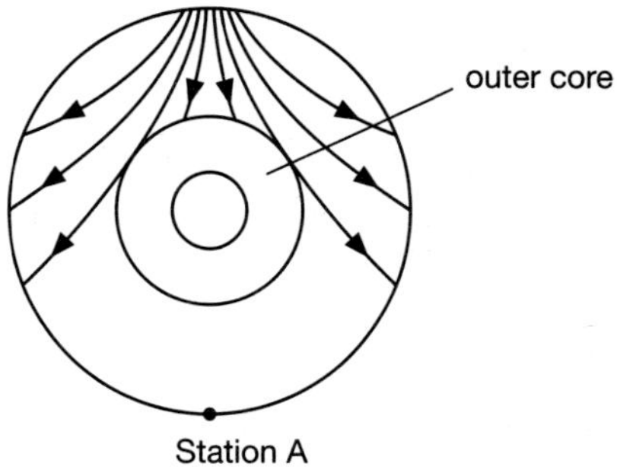

outer core

Station A

The results from these tests help scientists find out
about the structure of the Earth. Explain how.

_____

_____

_____

_____ [6 marks]

Answers are given on p.126

Exam Question and Answer

1) a) The graph shows the orbit time and average distance from the Sun of some planets and some asteroids. Asteroids are sometimes called minor planets.

Suggest why Jupiter takes longer than Mars to orbit the Sun.

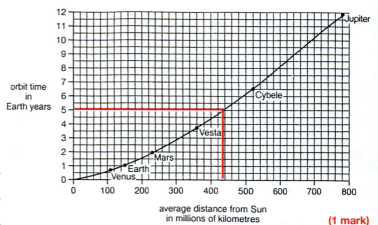

It has a greater distance to travel around its larger orbit **(1 mark)** as it is further from the Sun **(1 mark)**. [2 marks]

(1 mark)

b) The asteroid Ida is in orbit at an average distance of 430 million km from the Sun. Use the graph to find out how long it takes to orbit the Sun.

You **must** show clearly, **on the graph**, how you got your answer.

5 years **(1 mark)** [2 marks]

c) The orbit of Ida about the Sun is, in fact, elliptical. This means its speed varies during the orbit, like a comet.

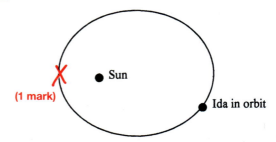

(1 mark)

i) Write an X on the diagram to mark the place where Ida will be travelling at its highest speed. [1 mark]

ii) Explain why it will be travelling at its highest speed at this place.

This is where the force of gravity is greatest, producing the greatest acceleration on Ida **(1 mark)**. [1 mark]

2) Explain the life cycle of a typical star like our Sun.

A huge cloud of dust and gas is pulled together by gravity **(1 mark)** causing high pressures or temperatures **(1 mark)**. At very high temperatures hydrogen nuclei undergo nuclear fusion **(1 mark)**. The star grows to become a red giant **(1 mark)**. It then shrinks to become a white dwarf **(1 mark)**. [5 marks]

3) The moon is a natural satellite of Earth. Artificial satellites are sent up by humans. Name four uses of artificial satellites.

Communications such as TV or phone **(1 mark)**.

Land watching (e.g. agricultural) or spying on military details **(1 mark)**.

Weather monitoring **(1 mark)**.

The Hubble telescope for space research **(1 mark)**. [4 marks]

How to score full marks

1 a) It would be unfair to ask you detailed facts about the solar system as they are too numerous. **Examiners will very often give you facts and ask you to explain them.** This question relies on the basic idea that planets further away from the Sun have a wider orbit and a greater distance to travel on each orbit.

b) The question asks you to use the graph to work something out. **You must do both tasks to get the chance of both marks.**

c) At the point where Ida is nearest the Sun, if it was travelling slower it would fall into the Sun.

2) **The life cycle of a star is a favourite question for many examiners.** This is a 5-mark question so you need to make 5 good scientific points.

3) **Satellites commonly appear in examination papers.** You will need to know their uses, which can be divided into the four categories indicated. Any valid technological use will gain the fourth mark. For some questions you will need to know about their orbits and explain circular motion.

Don't make these mistakes...

When explaining the uses of satellites try not to repeat the same idea. For example TV, phones and radio would only score 1 mark altogether because they are all examples of the same idea. Spying, weather, TV and space research would score 4 marks in total.

When writing about the life cycle of stars, try hard to get them in the correct order and be sure not to mix up the ideas. For a very large star, after the red giant phase they can start to glow brightly again as they undergo more fusion. In time, after expanding and contracting repeatedly, they explode in a supernova. These can form neutron stars and then even black holes, or instead they may form a second generation star like our Sun with its own solar system.

Key points to remember

Stars are much bigger and hotter than planets. They give out their own light and are often part of huge clusters called galaxies. Our Sun is a star and has planets orbiting it. The Sun is one of many millions in our galaxy, which is called the Milky Way. In the universe there are more than a billion galaxies.

Gravity keeps everything in orbit. There are satellites that orbit planets. Some are natural and called moons. Others are artificial and are used for four main things: communications, such as TV or phone, land watching (e.g. agricultural) or spying on military detail, weather monitoring, and such things as the Hubble telescope for space research.

Geostationary satellites have a high orbit and stay over the same part of the Earth as it rotates. They rotate with the Earth. Other satellites move around the outside of the Earth every few hours. Weather satellites need to orbit low to get good pictures. They travel in low polar orbits.

Gravity is an attraction between masses. Large masses like stars and planets have a stronger gravity, and masses weigh more near to them. Gravity decreases with distance from a planet. This is called the inverse square law. If you double the distance from a planet its weight will get less: four times less. If you get three times closer to the Sun, not only will you get very hot but you will get heavier: nine times heavier.

Earth has an atmosphere, and so when things fall weight accelerates them but drag tries to slow them down. When drag = weight there is no net force and no acceleration. The falling object has reached terminal velocity or terminal speed.

Days are caused by the rotation of a planet. The Earth rotates once every 24 hours and so we get one sunset each 24 hours.

The Earth orbits the Sun in $365\frac{1}{4}$ days. This is 1 year. Every 4 years we have a leap year (366 days) to make up for the $4 \times \frac{1}{4}$ days.

The seasons, spring, summer, autumn and winter, are caused by the angle of the Sun as the Earth is tilted in orbit. At low angles, the Sun's energy is diluted over the surface, causing less heating effect. When it is overhead it can be very hot indeed.

The Big Bang Theory says that the universe is expanding. There are three main groups of evidence for this: (1) light from distant stars and galaxies is red-shifted (i.e. they are moving away quickly); (2) light from extremely distant stars and galaxies is red-shifted even more (i.e. they are moving away very quickly indeed); (3) low frequency radiation seems to come from all parts of the universe.

A light year is the distance travelled by light in 1 year. Our nearest star is about 4.5 light years away and light travels at 3 000 000 000 m/s. Our nearest star does not seem very near at all!

Weight, mass and gravity are all connected:

Force = mass × acceleration

So weight (N) = mass (kg) × g (acceleration due to gravity)

On Earth g = $10 \, \text{m/s}^2$ so a 55 kg woman weighs $55 \times 10 = 550 \, \text{N}$.

On another planet g = $2 \, \text{m/s}^2$ so a 55 kg woman weighs only $55 \times 2 = 110 \, \text{N}$. Her mass has not changed but her weight has.

Planets reflect the Sun's light and we sometimes see them moving across the sky differently to stars. Planets orbit stars. Our planets orbit our Sun.

Questions to try

Examiner's Hint
● You will need to know about orbits, the position of the planets and how satellites behave before you tackle this question.

The diagram shows the orbit of Jupiter as it moves in our solar system.

Add to the diagram to show:

  (i) the position of the **Sun**, label it S   [1 mark]

  (ii) the orbit of **Mercury**, label it M   [1 mark]

(iii) the orbit of **Uranus**, label it U.   [1 mark]

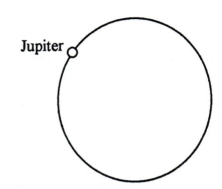

Jupiter

Examiner's Hint
● Be sure you understand about Earth's day and night and its seasons.

(a) TV satellites have geostationary orbits. Explain what a geostationary orbit is and why these orbits are particularly useful for TV satellites.

_____

_____

_____ [3 marks]

(b) Weather satellites are not in geostationary orbits. Suggest why.

_____

_____ [2 marks]

(c) What is the name of the Earth's natural satellite?

_____ [1 mark]

(d) The Earth has day and night every 24 hours. Explain why.

_____

_____ [2 marks]

(e) Great Britain has four seasons in the year. Winter is cooler than summer. Explain why.

_____

_____

_____ [3 marks]

Examiner's Hint
● You need to learn about the types of stars and their life cycles.

(a) Most scientists believe the Universe is expanding.
What is this theory called and what evidence supports it?

_____

_____

_____ [4 marks]

(b) Scientists use light years to help them describe space. What is a light year?

_____

_____ [3 marks]

(c) Stars have a **life cycle**. There are **two types** of stars:
big first generation stars and second generation stars like our Sun.

(i) Explain the life cycle of a **second generation** star.

_____

_____

_____

_____ [6 marks]

(ii) **A first generation star also has a life cycle.** Explain how it is different
from the life cycle of a **second generation** star.

_____

_____ [3 marks]

Answers are given on p.127

20 Radioactivity

Exam Question and Answer

1) a) i) Explain why some substances are radioactive and others are not.

> An element has different isotopes (different numbers of neutrons but the same number of protons). Radioactive isotopes are radioactive and their nuclei are unstable **(1 mark)**. They undergo radioactive decay and emit gamma radiation or radioactive particles (beta or alpha). This can then make their nuclei more stable **(1 mark)**. [2 marks]

ii) State a cause of background radiation.

> Background radiation comes from rocks in the Earth **(1 mark)**. [1 mark]

iii) Explain what you understand by the meaning of the **half-life** of a radioactive element.

> Half-life is the time it takes for half **(1 mark)** of the radioactive atoms now present to decay **(1 mark)**. [2 marks]

b) Technetium 99 is a radioactive material with a half-life of 6 days. It is used to study blood flow around the body. A sample of technetium 99 has an activity of 96 counts per minute when injected into a patient's blood stream. Estimate

i) its activity after 12 days

> Activity after 0 half-lives (0 days) = 96 counts per minute
> Activity after 1 half-lives (6 days) = 48 counts per minute
> Activity after 2 half-lives (12 days) = 24 counts per minute **(1 mark)**
>
> [1 mark]

ii) how long it will take for the radioactivity from the injection to become undetectable

> Its activity will probably never fall to zero but after 3 half-lives (18 days) it will fall to 12 counts per minute and this is about the same as background radiation **(1 mark)**. [1 mark]

c) Technetium 99 is a gamma (γ) emitter and does not produce alpha (α) or beta (β) radiation. Explain why it is safe to inject technetium 99 into the body.

> Gamma radiation is highly penetrative and will leave the body **(1 mark)**. Its ionising power is not all used to destroy human cells in the body as some remains when it leaves **(1 mark)**. [2 marks]

2) Radioactive carbon 14 emits a β **(beta) particle** when it decays.
It has a **half-life** of 5600 years.

a) Explain what you understand by the terms in bold print.

A beta particle is a negative electron that is emitted from the
nucleus **(1 mark)**. Half-life is the time it takes for half **(1 mark)** of the
radioactive atoms now present to decay **(1 mark)**.

[3 marks]

How to score full marks

1) There is often a lot of information in questions on radioactivity. You need to **use the information** and write **clear scientific explanations** or make **good scientific decisions**. Many of these questions rely on the key ideas of **half-life**, **penetration of radiation**, **ionising power** and **background radiation**. Often it is not obvious from the question which key ideas you should use. The examiners are asking you to **prove your understanding** of the key ideas **by applying them** to things you may not have seen even before you entered your examination. **Don't be put off**; just **pause** a little, **don't panic**, read the question again and **use your understanding to make sense of the question**.

1) iii) You can also describe half-life as the time it takes for a radioactive material to become half as radioactive.

2) **Beta particles are electrons emitted from the nucleus.** They originate from neutrons which are neutral because they have one proton+ and one electron– combined. When the electron is emitted as a beta particle it loses one negative charge and becomes a proton.

Don't make these mistakes...

When defining half-life, avoid writing about atoms and instead write about activity.

Half-life calculations often seem so easy that you feel like doing them in your head. It is often better to write it down in steps so that you do not make a simple error.

Gamma is more penetrative than beta. Beta is more penetrative than alpha. But it is not true that alpha is the least dangerous to humans. Other factors need to be considered.

Key points to remember

Atoms are neutral and the charges balance. The number of positive protons = the number of negative electrons. When electrons are added or removed from an atom this balance changes and the atom changes to an ion.

Background radiation mainly comes from rocks, radon and thoron gas in the air and cosmic rays. It is always there and when doing an experiment you may need to subtract the background count in order to find the count rate of a material.

Particle	Charge	Mass
Electron	–1	1/2000 (i.e. 2000 time **less** massive than a proton)
Proton	+1	1
Neutron	0	1

The nucleus of an atom has protons and neutrons. Although most of the mass is in the nucleus it is very small indeed. The even smaller and tinier negative electrons move around the outside of the nucleus covering lots of space. This marks the overall size of the atom but even then it is mostly space!

Emissions from tracers in medicine, industry and the environment can be tracked by using radioactivity. Radiation kills living cells and can be used to sterilise food and medical instruments as well as curing some cancers. Radiation penetrates matter to different extents and can be used to gauge and monitor the thickness of materials.

Half-life is the time it takes for half of the radioactive atoms now present to decay. Or to put it another way, half-life is the time it takes for a radioactive material to become half as radioactive. The half-life of carbon 14 can be used to find out how long ago living things died.

Isotopes of an element have the same atomic number but different mass numbers. The number of neutrons are different. This can cause the nucleus to be unstable and radioactive, causing particles to be emitted. Two isotopes of carbon are carbon 12 and carbon 14.

Carbon 12 is stable and not radioactive $^{12}_{6}C$

Carbon 14 is unstable and is radioactive $^{14}_{6}C$

Geiger counters are used to detect radioactivity

Radiation can be detected by ionisation. Geiger–Muller tubes, spark counters and photographic film are sensitive to radiation because it ionises matter.

Mass number is the:

number of protons + number of neutrons

Atomic number is the number of neutrons

$^{12}_{6}C$ is carbon 12

Radiation ionises atoms. It can destroy or damage living cells and even cause cancer. It can be used to kill unhealthy cells such as some cancerous ones.

The three types of radiation are alpha (α), beta (β) and gamma (γ).

Radiation	Mass	Nature	Charge	Blocked by
Alpha (α)	4	helium nucleus	+2	a few sheets of paper
Beta (β)	1/2000	electron from nucleus	–1	a few cm of aluminium
Gamma (γ)	0	electromagnetic wave	0	thick lead

Questions to try

Q1-2

Examiner's Hints
- You need to know about half-life and how different types of radiation penetrate materials and can be used to measure their thicknesses.
- Be prepared with a pencil to collect easy marks by plotting a graph.
- Be sure to know about the different types of radiation, their nature and their properties.

(1) Kate's teacher wants to find out how much beta radiation passes through different thicknesses of aluminium.

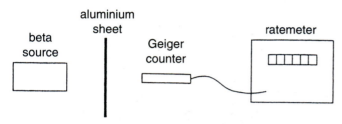

First she measures background radiation. It gives a reading of 60 counts per minute on the ratemeter. She now records the count rate for different thicknesses of aluminium. The table shows the results.

Thickness of aluminium in mm	1.0	2.0	3.0	4.0	5.0	6.0	7.0	8.0
Actual ratemeter reading in counts per minute	560	310	180	120	90	75	60	60
Corrected count rate in counts per minute	500	250						

(a) Finish the table. There are 6 gaps. [1 mark]

(b) Plot the points on the grid. [1 mark]

(c) Finish the graph by drawing the best curve. [1 mark]

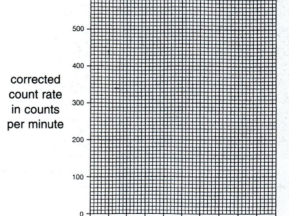

corrected count rate in counts per minute

thickness of aluminium in mm

(d) Aluminium is rolled into sheets 20 mm thick in a rolling mill.

A radioactive source and a detector are used to check the thickness of the sheet as it leaves the rollers.

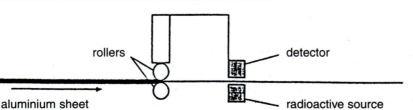

(i) Why is beta radiation **not** suitable for checking a 20 mm sheet?

_____ [1 mark]

(ii) Suggest one type of radiation which could be used to check the thickness of a 20 mm sheet.

_____ [1 mark]

(iii) The radioactive isotope has a half-life of 5.3 years. Explain what is meant by the term **half-life**.

_____

_____ [2 marks]

(iv) One rolling mill uses 20 mg of this isotope as the source. What will be the mass of this radioactive isotope in the source 10.6 years later?

Mass = _____ mg   [1 mark]

(v) It would not be sensible to use, in the rolling mill, a radioactive isotope with half-life much shorter than 5.3 years. Explain why.

_____

_____ [2 marks]

(2) There are **three** types of nuclear radiation.
(a) Complete the following sentence.

The **three** types of radiation are alpha, and gamma.   [1 mark]

(b) Complete the table.

Type of radiation	Charge	Nature	Penetrating power
alpha	+2		stopped by paper
gamma		electromagnetic wave	absorbed by thick lead/ concrete

[2 marks]

(c) What does ionising mean?

_____

_____ [2 marks]

Answers are given on p.128

1 Life Processes and Cells

Q1 How to score full marks

Any **four** of the following ideas for 1 mark each:

- sugar has dissolved in moisture (on the surface of the fruit) (**1 mark**)
- this solution is more concentrated than the solution inside the fruit (**1 mark**)
- osmosis / diffusion (**1 mark**)
- movement of water out of fruit (**1 mark**)
- through partially permeable membrane (of fruit cells) (**1 mark**)

🎯 **To get full marks you must answer fully and clearly. Good exam candidates can easily miss out on some of these points because they think they are too obvious to need mentioning. Use your answer to show just how much you do know and understand.**

Q2 How to score full marks

(a) Any **four** of the following ideas for 1 mark each:

- more respiration (**1 mark**)
- carbon dioxide levels increase (**1 mark**)
- (more) carbon dioxide in blood (**1 mark**)
- pH of blood increases (**1 mark**)
- detected by brain / medulla (**1 mark**)
- (and) detected by (chemo) receptors in (aortic / carotid) arteries (**1 mark**)
- signals sent to diaphragm / intercostal muscles (**1 mark**)

🎯 **Questions about respiration are often set in some context, in this case exercise. You have been given some information in the question. You will not get marks for repeating any of those points. This is a clue that to gain full marks you have to describe the changes that follow on from an increase in aerobic respiration.**

(b) (i) cannot breathe fast enough / circulation not fast enough (**1 mark**)

🎯 **This is more straight forward than (a). Don't assume that the later parts of questions will always be harder.**

(ii) glucose / sugar / $C_6H_{12}O_6$ (**1 mark**) → lactic acid (**1 mark**)

🎯 **A correct chemical formula may gain a mark in this question, but it is best to give words when you are asked for a word equation. If you are asked to give chemical formulae then do so then or you will not gain marks.**

(iii) Any **one** of the following ideas for 1 mark:

- produces less energy (**1 mark**)
- less efficient (**1 mark**)
- produces lactic acid (**1 mark**)
- does not produce water (**1 mark**)
- does not produce carbon dioxide (**1 mark**)
- produces an oxygen debt (**1 mark**)

🎯 **You will not gain a mark for stating that anaerobic respiration does not use oxygen. This is correct but you have already been told this in the question.**

Q1 How to score full marks

(a) Any **four** of the following ideas for 1 mark each:

- digested / broken down / made soluble (**1 mark**)
- by protease / enzyme (**1 mark**)
- in stomach / in small intestine / from stomach / from pancreas (**1 mark**)
- into amino acids (**1 mark**)
- amino acids / small molecules absorbed into blood (**1 mark**)

Although you may not need all the space available to write enough to get full marks, the number of lines allowed is indicating that a detailed answer is required. Because you have not been told in the question to write about digestion or the role of enzymes, for example, then you can be awarded marks for these points. Don't miss them out because you think they are obvious.

(b) Any **three** of the following ideas for 1 mark each:

- lipase / enzyme works best in alkaline / neutral conditions (**1 mark**)
- acid denatures or inactivates enzyme / inhibits enzyme activity (**1 mark**)
- bile emulsifies fat / bile produces larger surface area of fats / bile alkaline (**1 mark**)
- for enzyme to work on / which increase activity of enzymes (**1 mark**)

Use the information in the table to comment on the effect of the different conditions, but then use your own knowledge to expand on this. Remember that enzymes are affected by conditions such as temperature or, in this case, pH, and that each enzyme has its own optimum conditions. In questions about fat digestion remember that bile helps by emulsifying the fat droplets (breaking large droplets into smaller ones) as well as being alkaline.

Q2 How to score full marks

(a) bronchus (**1 mark**) (top line)
bronchiole (**1 mark**)

You should have learnt these. If you are not sure in an exam, then think about which items on the list cannot be correct answers and choose from the rest.

(b) Any **four** of the following ideas for 1 mark each:

- rib cage moves up / out or intercostal muscles contract (**1 mark**)
- diaphragm moves down / contracts / flattens (**1 mark**)
- increase volume / size (**1 mark**)
- pressure drops (**1 mark**)
- pressure differential / external air pushed in (**1 mark**)

Not air sucked in or pulled in.

Air enters and leaves the lungs during breathing because of *pressure* changes. These are caused by changes in the *volume* of the lungs. If the pressure inside the lungs is less than the air pressure outside then air will enter. Do *not* talk about air being 'sucked' in.

Q1 How to score full marks

(a) arrow from sensory to motor neurone **(1 mark)**

🎯 **In a nerve pathway signals pass along sensory neurones (from sense organs) before they pass along motor neurones and cause a response.**

(b) synapse **(1 mark)**

🎯 **Make sure you learn scientific names.**

(c) chemical / named transmitter **(1 mark)** diffusion **(1 mark)**

🎯 **You are not expected to know the names of transmitter chemicals but of course you will get the mark if you do. Remember that they move by diffusion after being released from the first neurone.**

(d) nicotine / anaesthetic / alcohol / solvents / ecstasy / adrenaline / paracetamol / curare / caffeine **(1 mark)**

🎯 **The question asked you to 'suggest'. All of the chemicals above affect our nervous system in some way and are therefore valid suggestions.**

(e)(i) receptor **(1 mark)** effector **(1 mark)**

🎯 **You could also gain marks by giving examples of receptors (e.g. eye, ear or other sense organs) or effectors (e.g. muscles or glands).**

(ii) fast / protection / no thought / involuntary **(1 mark)**

🎯 **Reflexes happen quickly because they don't involve conscious thought. Speed may be important if the reflex is in response to something that may harm the body.**

Q2 How to score full marks

(a)(i) progesterone **(1 mark)**

🎯 **Progesterone maintains the uterus (womb) lining. When the levels drop the lining breaks down and the menstrual bleeding (period) starts. This matches up with the graph for X.**

(ii) oestrogen **(1 mark)**

🎯 **Increasing amounts of oestrogen cause the uterus lining to thicken. This matches up with the graph for Y.**

(b) ovulation / prevents ovulation **(1 mark)**

🎯 **High levels of oestrogen trigger off ovulation (egg release from the ovaries). High levels of progesterone prevent ovulation.**

(c) contraceptive / hormone replacement therapy / to stabilise periods / treat infertility / move periods / post menopausal pregnancies **(1 mark)**

🎯 **You should know an answer to this already but you could use information from the graph to work out that you could use hormones to control periods.**

(d) Any **four** of the following for 1 mark each:
idea of insulin responsible **(1 mark)**
released when blood sugar / glucose rises **(1 mark)**
glucose converted to glycogen **(1 mark)**
in liver / muscles **(1 mark)**
less insulin when glucose drops **(1 mark)**
insulin facilitates cell uptake of insulin **(1 mark)**

🎯 **The instruction to 'write about' is an extra prompt for you to write at some length giving as much detail as you can.**

4 Plants

Q1 How to score full marks

(a) auxin / IAA / indole acetic acid (**1 mark**)

🎯 **You may have come across the hormone under these different names.**

(b)(i) cannot detect light / has not responded to light (**1 mark**)

auxin / hormone evenly distributed (**1 mark**)

🎯 **You are asked to 'explain' the results so you will get no marks if you simply describe what has happened, e.g. by saying 'the seedling has grown taller'.**

The first mark is for explaining the significance of the foil and the second for explaining how this affects the hormone. Without the foil the hormone would have accumulated on the shaded side, causing that side to elongate, making the tip bend towards the light.

(ii) auxin / hormone not made / made in tip / no auxin / hormone (**1 mark**)

🎯 **The hormone is made in the tip so with the tip removed no hormone is produced. No hormone, no elongation.**

Q2 How to score full marks

(a) temperature / low temperature (**1 mark**)

🎯 **The graph shows that with the same light intensities and the same high carbon dioxide concentration the rate of photosynthesis will increase if the temperature is higher.**

(b) high, high, high (**1 mark**)

🎯 **As this is a very straightforward question all 3 ticks need to be correct for 1 mark. In this type of question if you give too many ticks it shows a lack of understanding and will lose the mark(s).**

(c) increases carbon dioxide and/or temperature (**1 mark**) so more photosynthesis / more growth / more enzyme activity (**1 mark**)

🎯 **You won't get marks for an answer like 'the lettuce will grow bigger' as this is just a rewording of the fact that the yield will increase, which you've just been told.**

(d) whether cost of paraffin (**1 mark**) is less than profit from increased yield / if there is a market for increased lettuces (**1 mark**)

🎯 **This question is not really testing your scientific knowledge but it is about the real life application of scientific ideas, so expect to get the odd question like this.**

5 Ecology and the Environment

Q1 How to score full marks

(a) (X is) as food / being eaten / nutrition / consumed (**1 mark**)

(Y is) photosynthesis (**1 mark**)

(Z is) respiration (**1 mark**)

The carbon (and nitrogen cycle) can be written in different ways but the processes will always be the same. Make sure you look at the direction of the arrows to help you answer.

(b)(i) conversion of ammonia / ammonium compounds / nitrites / formation of nitrites / nitrates (**1 mark**)

In this case you can gain the mark by correctly stating either what the bacteria use up or what they produce. In some questions you might have to give both. So if you are able, give as full an answer as you can.

(ii) conversion of ammonia / ammonium compounds / nitrates / formation of nitrogen (gas) (**1 mark**)

Note that denitrifying do *not* do the opposite of nitrifying bacteria.

Q2 How to score full marks

(a) W (**1 mark**)

The oak trees can support many animals because they are so big but there is only a very small number of them compared with the animals feeding off them.

(b) consumers eat / unable to make food (**1 mark**)

(eat) primary consumers / herbivores (**1 mark**)

The phrase 'write about' is to encourage you to give sufficient details to give you a good chance of getting both points, although you could get both marks with a relatively short answer.

(c)(i) average mass or weight of each feeding level/ **or** total mass or weight of each feeding level/ **or** mass of all / each organism(s) in the table (**1 mark**)

Strictly speaking, biomass is the dry weight of the living thing, i.e. the weight apart from any water contained, but in practice this can be difficult to measure so mass or weight can be used.

(ii) energy lost / used up (moving up chain) (**1 mark**) plus one of: not all parts eaten / not all parts digested / respiration / heat / movement / muscle activity / chemical reactions in cells / excretion (**1 mark**)

A common mistake here would be to say that the oak trees are larger. You need to explain why there is less *overall* biomass as you go up the pyramid.

Q1 How to score full marks

(a) Any **two** of the following ideas for 1 mark each:
- mixture of genes / alleles / chromosomes / information / characteristics / gene shuffling **(1 mark)**
- mutations **(1 mark)**
- environmental reasons / examples (diet / damage / disease / activity) **(1 mark)**

Two environmental reasons can be used for 2 marks.

Although this question is about human babies it is really only asking the basic question of why living things show variation.

(b) 46 / 23 pairs **(1 mark)**

23 **(1 mark)**

Virtually all human cells contain 46 chromosomes. The main exceptions are the sex cells, eggs and sperm, which have half of this. This means that when they join together in fertilisation, the fertilised egg cell will have the full number, 46.

(c) sex cells: X (X) and X, Y **(1 mark)** (these need to be shown separated)

children: correct derivation of XX (girl / Sarah) and XY (boy / James) **(1 mark)**

You can either set out your answer as in the question on page 34, or use a Punnett square. Either way you must make it clear which of your answers shows the sex cells and which shows the children.

		Male gametes	
		X	Y
Female gametes	X	XX (girl)	XY (boy)
	X	XX (girl)	XY (boy)

Q2 How to score full marks

(a) Any **two** of the following ideas for 1 mark each:
- fossilisation is rare / they did not live in conditions favourable for fossilisation / did not fossilise **(1 mark)**
- fossils not yet found **(1 mark)**
- evolution in rapid bursts **(1 mark)**
- fossils destroyed **(1 mark)**
- evolution may not have occurred **(1 mark)**

There are lots of reasons why the fossil record might be incomplete. The word 'suggest' shows that you are only being asked to come up with *possible* reasons.

(b) not all body parts fossilised / not all parts of fossil found / fossil damaged or distorted **(1 mark)**

Don't try to repeat information from (a) by saying that we don't have the fossils. As this is a different question it will be about a different idea.

(c) Any **four** of the following ideas for 1 mark each:
- variation in brain size **(1 mark)**
- (because) larger brain increases intelligence **(1 mark)**
- (larger brain) increases chances of survival / producing more offspring **(1 mark)**
- brain size is controlled by genes / larger brain size is inherited / passed to children **(1 mark)**
- repetition / over many generations **(1 mark)**
- changed gene pool / increased proportion of larger brains **(1 mark)**

The stages in the process of natural selection are the same what ever the example. This one happens to be about brain size, but the same ideas would apply to any other example. The only difference would be that in another question you would have to explain why the different feature would be an advantage.

Q1 How to score full marks

(a) (i) NH_4^+ / SO_4^{2-} (**1 mark**)

🎯 **If you have to select from a list of formulae, spend time to check that you have copied down the formula as it was written in the list.**

(ii) Na^+ / NH_4^+ (**1 mark**)

🎯 **Cations are positive ions so you must look for any particle with a positive charge. The question only asks for one cation: beware, if you give two answers they will both have to be correct to get the mark. Remember there are no extra marks to be awarded; in this case the best advice is to give only one answer.**

(iii) Na_2CO_3 (**1 mark**)

🎯 **You should recognise the carbonate group from the formula given and use this to select the correct formula.**

(iv) H_2SO_4 (**1 mark**)

🎯 **This one of the formulae you should remember.**

(v) NH_3 (**1 mark**)

🎯 **This is another formula you should know. Remember, ammonia is a covalent compound and ammonium is a cation. Do not confuse the two.**

(b) $(NH_4)_2SO_4$ (**1 mark**)

🎯 **You should be able to work out the formula by using the constituent ions given in the list, NH_4^+ and SO_4^{2-}.**

Q2 How to score full marks

(a) (i) barium + water → barium hydroxide + hydrogen (**1 mark**)

🎯 **You would be allowed to put an equals sign instead of the arrow. Do not include heat in the equation although you could still use ΔH = –ve as in the symbol equation.**

(a) (ii) fizzes / makes a colourless gas (**1 mark**)

barium dissolves / colourless solution (**1 mark**)

solution gets hot / temperature increases (**1 mark**)

🎯 **You can work out a lot of information from a chemical equation. In this case the state symbols of the products should enable you to work out that a solution is made, the (aq), and that a gas is produced, the (g). In addition, because the ΔH value is negative this tells you it is an exothermic reaction so the temperature of the reaction mixture should rise.**

(b) (i) $Ba(NO_3)_2$ (**1 mark**)

🎯 **You should use the charge on the barium ion and on the nitrate ion to work out the formula.**

(ii) BaO (**1 mark**)

🎯 **You should remember that an oxide ion has the formula O^{2-}.**

(iii) $BaCl_2$ (**1 mark**)

🎯 **You should remember that a chloride ion has the formula Cl^-.**

8 Structure and Bonding

Q1 How to score full marks

(a)(i) NaH **(1 mark)**

🎯 You need to remember that metals form positive ions and in the group 1 elements the charge on the ion is +1. So to make certain that the charges cancel out the formula must be Na^+H^-.

(ii) gain of 1 electron **(1 mark)**

🎯 Remember that an electron is negative, so to make a negative ion electrons must be gained. In this question it is important you show the examiner that you know how many electrons a hydrogen atom gains.

(b)(i) drawing showing electron structure of 2.8.1 **(1 mark)**

🎯 In an exam drawing circles free hand will do. You will be expected to be able to work out the electron structure for any of the first 20 elements. Because it is group 1, then sodium must have one electron in its outer shell.

(ii)

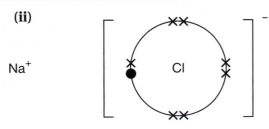

Na^+

correct electron structure for Na^+ (2.8) **(1 mark)**

correct electron structure for Cl^- (2.8.8) **(1 mark)**

correct charge on the ions **(1 mark)**

🎯 Notice that on the 'dot and cross' diagram only the outer electrons have been drawn. Unless it is stated in the question that you have to draw all the electrons, this is an accepted way of completing 'dot and cross' diagrams. You should also notice that the drawing shows the charges on the ions and that the drawing cannot be confused with covalent bonding because the two ions are deliberately drawn away from each other.

(iii) in solid sodium chloride ions are fixed in position **(1 mark)**

in molten sodium chloride ions are free to move **(1 mark)**

🎯 You should notice that two statements are made, one about the solid and one about the molten liquid. This should guarantee you cover both marks available.

(c)(i) covalent **(1 mark)**

because they are both non-metals / both atoms need to gain one electron to get a stable set of electrons in the outer shell **(1 mark)**

🎯 It is easier to understand bonding if you remember that compounds containing metals are ionic and those with only non-metals are covalent. If this was hydrochloric acid, the bonding would be ionic because a hydrogen atom loses one electron and the chlorine atom gains one electron, so both atoms get a stable electron structure. Hydrogen chloride should really be written HCl(g) and hydrochloric acid HCl(aq).

(ii)

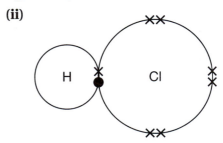

one shared pair between the hydrogen atom and chlorine atom **(1 mark)**
rest of the structure correct **(1 mark)**

🎯 Notice that only the outer electrons have been drawn. The shared pair is clearly shown in the diagram. This time it is important that the two outer shells drawn actually touch.

(iii) $H_2 + Cl_2 \rightarrow 2HCl$ **(1 mark)**

🎯 This is a straightforward chemical equation. You should remember that hydrogen and chlorine are diatomic molecules.

(iv) low melting point / low boiling point **(1 mark)**

does not conduct electricity **(1 mark)**

🎯 This question is not about hydrogen chloride but about compounds with simple molecular structures. All simple molecular structures have certain properties in common. Do not include low melting point and low boiling point as your two examples, because they are similar points and you will only get 1 mark.

9 Fuels and Energy

Q1 How to score full marks

(a) fermentation of glucose solution takes a few days to be complete (**1 mark**)

so that ethanol burnt can be replaced straight away (**1 mark**)

🎯 **You must not write that ethanol can be reused. Once a fuel has been burned it is changed into carbon dioxide and water so it cannot be used again. To be renewable the fuel must be able to be made in a very short period of time to replace the fuel burned.**

(b)(i) compound containing carbon and hydrogen (**1 mark**) only (**1 mark**)

🎯 **You must be careful to make certain you do not refer to a mixture of hydrogen and carbon. Carbon and hydrogen are chemically bonded together in a hydrocarbon. You must also make it clear in your answer that no other element is present in a hydrocarbon.**

(ii) it contains oxygen (**1 mark**)

🎯 **The chemical formula has the information you need.**

(c) suitable container to burn fuel, e.g. a spirit burner or a metal crucible (**1 mark**)

use of one gram of fuel (**1 mark**)

suitable container, e.g. a copper can, containing water above the burner (**1 mark**)

same amount of water (**1 mark**)

measure the temperature change of the water (**1 mark**)

largest temperature change indicates fuel that gives out the most energy (**1 mark**)

🎯 **It is easier to answer this question with the aid of a large detailed labelled diagram. Make certain that the quantities used and the measurements that need to be taken are also included on the diagram.**

Another way of answering the question could involve measuring the mass of burner before and after the experiment, using a known amount of water and using the relationship energy = mc Δ T, if necessary accounting for different masses of fuel being burned.

(d)(i) Any **two** from:
carbon monoxide (**1 mark**)
carbon (**1 mark**)
water (**1 mark**)

🎯 **You should realise that water is formed during complete and incomplete combustion of methanol.**

(ii) Bond breaking absorbs energy from the surroundings (**1 mark**). Bond making releases energy into the surroundings (**1 mark**). More energy is released than absorbed (**1 mark**).

🎯 **It is easier to answer a question like this with 3 distinct sentences. This forces you to make the 3 required marking points. Avoid using phrases such as 'bond making needs more than bond breaking'. This does not describe the direction of energy flow, i.e. whether it is entering or leaving the surroundings. The more energy in the surroundings the hotter it becomes.**

Q2 How to score full marks

Any **three** from:

cracking involves the breakdown of large hydrocarbons (**1 mark**) into smaller ones (**1 mark**)

cracking makes smaller chain alkenes and alkanes (**1 mark**)

cracking involves breaking single bonds between carbon atoms (**1 mark**)

and

any **three** from:

polymerisation involves small molecules becoming bigger molecules (**1 mark**)

the small molecules are often alkenes (**1 mark**) and are called monomers (**1 mark**)

double bonds between carbon atoms are broken during polymerisation (**1 mark**)

🎯 **The mark allocation indicates that this needs an extended answer. You must make 6 distinct points, and because the question refers to two processes a sensible approach would be to write 3 points about cracking and 3 points about polymerisation. Make certain that you do not muddle the two processes. Write a paragraph about cracking and then another one about polymerisation.**

Q1 How to score full marks

(a) uses a lower temperature (**1 mark**)

less energy is needed to maintain the temperature and so it is cheaper to run (**1 mark**)

🎯 **It is important that you use the information in the text and make a sensible deduction from the data. Economics is a driving force for all industrial processes.**

(b) carbon (**1 mark**)

🎯 **Some things you just have to learn.**

(c) the ions (**1 mark**)
move (**1 mark**)

🎯 **You should know that for any material to conduct electricity there has to be a particle that can move and carry an electric charge. In an electrolyte the ions are free to move because it is a liquid. As an alternative you could have referred to the movement of one of the ions referred to in the question.**

(d) oxygen (**1 mark**)

🎯 **You can work out this answer because when simple ions react at an electrode they make atoms and sometimes later on molecules. Oxide ions can only make oxygen so this must be the gas. Don't attempt to write a formula when a name will do: remember if a formula is used it has to be totally correct.**

(e) $Al^{3+} + 3e^- \rightarrow Al$
correct reactants and products (**1 mark**)
balanced (**1 mark**)
it is reduction because electrons are gained by the aluminium ion (**1 mark**)

🎯 **Electrode or half equations are very difficult, but if you are careful then you can guarantee most of the marks. Remember that the reactant, in this case the aluminium ion, goes on the left and the product goes on the right. With ionic equations both the charge and the symbols need to be balanced. It is easy to balance the symbols but to balance the charge you have to use electrons. Remember that e⁻ is the symbol for an electron.**

OIL RIG is a good way of remembering oxidation and reduction. *O*xidation *I*s *L*oss and *R*eduction *I*s *G*ain of electrons.

Q2 How to score full marks

Tectonic plates move because they are slowly dragged along by movement of material within the mantle (**1 mark**).

In mountain building two plates move towards each other (**1 mark**).

The plates buckle upwards as they move closer and closer to each other, and this forces the crust upwards and a mountain is formed (**1 mark**).

In the mid-Atlantic oceans two plates are moving apart (**1 mark**).

Molten magma escapes and solidifies as new igneous rock (**1 mark**).

As the plates keep moving, more and more new igneous rocks are formed and the original plate boundaries move further and further apart (**1 mark**).

🎯 **One misconception that many students have is that the tectonic plates are just part of the crust. In fact you should realise that they include part of the upper mantle as well.**

You could have drawn two diagrams, and with appropriate labels the explanation is much easier to make than with words alone. Make certain you show that the tectonic plates can move towards each other and away from each other. There is no need to refer to continental and oceanic plates, but if you do you must do so correctly. The mark allocation indicates that 6 clear points must be made, and you must be careful with the diagrams that 6 points have been made.

11 Chemical Reactions

Q1 How to score full marks

(a) zinc + hydrochloric acid → hydrogen + zinc chloride **(1 mark)**

🎯 If a question asks for a word equation do not give a symbol equation; you must show the examiner you know the difference between a word equation and a symbol equation. You can use an equals sign instead of an arrow.

(b)(i) final volume is the same as the original graph: $185\,cm^3$ **(1 mark)**

same shape as original graph but always on the left of it **(1 mark)**

🎯 Remember that the temperature will only change the rate of reaction, giving a steeper gradient to the graph at the start, but will not change the volume of product made. The volume of hydrogen will only change if the quantity of reactants is changed, as in (b) (ii).

(ii) Half of the mass of zinc will give half of the original volume **(1 mark)**.

(c) The reacting particles are more crowded **(1 mark)** so there are more collisions per second **(1 mark)**.

🎯 As an alternative to collision frequency you can refer to the number of collisions per second. The question is worth 2 marks so the examiner is looking for 2 distinct points.

(d) The lumps have a smaller surface area **(1 mark)**. So there are fewer collisions per second **(1 mark)**.

🎯 You must show the examiner that you realise that the surface area of a lump is much smaller than that of a powder. Don't refer to the particle size of the zinc, only use the term particle to refer to ions, atoms and molecules, i.e. the reacting particles. Again, you will get the second mark if you make it clear you know that there will be less collisions per second.

(e) Copper powder is the catalyst **(1 mark)** because the copper does not change colour **(1 mark)** and the reaction is faster than the reaction between just zinc and hydrochloric acid **(1 mark)**.

🎯 The mark allocation gives a clue that there are 3 points to make. One mark is obviously for the choice of the catalyst, and the other 2 marks are for your explanation using the data in the table. You must make it clear that a catalyst makes a reaction faster, so the table included a control experiment, namely the zinc on its own. A catalyst is chemically unchanged at the end of a reaction, and this is indicated by the lack of colour change in the table.

Q1 How to score full marks

(a) they have 7 electrons in the outer shell (**1 mark**)

🎯 It is important that you refer to the outer shell in your answer. Remember that the group number is the same as the number of electrons in the outer shell.

(b) chlorine is 2,8,7 (**1 mark**)
so it has three occupied electron shells (**1 mark**)

🎯 The question is worth 2 marks so your answer must have 2 distinct points. Remember that all elements in the same period have the same number of occupied electron shells.

(c) (i) gains one electron (**1 mark**)

🎯 All atoms try to obtain a stable octet of outer electrons. So a chlorine atom needs one extra electron. It is not enough to state that chlorine gains electrons.

(ii) I^- (**1 mark**)

🎯 Remember that if a chlorine atom needs to gain one electron to form a chloride ion, then an iodine atom also needs to gain one electron. So the formula for an iodide ion shows one negative charge.

(d) (i) iron + chlorine $\rightarrow$ iron(III) chloride (**1 mark**)

🎯 The names of the reactants and the products were given in the question. Just remember to put the reactants on the left of the arrow and the products on the right of the arrow.

(ii) $2Fe + 3Br_2 \rightarrow 2FeBr_3$
correct formulae of reactants and products (**1 mark**)
correct balancing (**1 mark**)

🎯 You must remember that the elements in group 7 exist as diatomic molecules, so bromine has the formula Br_2. You can also balance this equation by using fractions:

$Fe + \frac{3}{2}Br_2 \rightarrow FeBr_3$

(iii) chlorine gains electrons more easily than bromine (**1 mark**)

🎯 The more reactive a group 7 element, the easier its atoms gain electrons. Do not refer to the electrons being gained faster; this is incorrect.

(e) $2KI + Cl_2 \rightarrow 2KCl + I_2$
correct formulae of reactants and products (**1 mark**)
correct balancing (**1 mark**)

🎯 You should remember that chlorine can displace bromine (in bromides) and iodine (from iodides) because it is more reactive than bromine and iodine. You should also remember the formula for alkali metal halides such as potassium iodide or potassium chloride.

(f) (i) black/dark (**1 mark**) solid (**1 mark**)

🎯 The description for bromine gives a hint that the answer should give a colour and the state of the element at room temperature. You are not expected to know the appearance of astatine but should be able to work out it from the change in appearance of chlorine to iodine.

(ii) no reaction (**1 mark**)

astatine is less reactive than iodine / iodine is more reactive than astatine (**1 mark**)

🎯 Do not be tricked into thinking that there has to be a reaction when two substances are mixed together. Astatine is the least reactive of the group 7 elements so it cannot displace the other elements from their compounds.

(g) neon has a stable outer octet of electrons (**1 mark**)

🎯 This question is not about fluorine but about neon. Remember the noble gases are unreactive elements. Do not just refer to a stable octet of electrons: remember, it is the outer electrons that are important in determining the reactivity of elements.

13 Chemical Calculations

Q1 How to score full marks

(a) 24 (**1 mark**)

(b) $24 \times 6 \times 10^{23}$ (**1 mark**)

1.44×10^{25} (**1 mark**)

🎯 **Remember, one mole contains 6×10^{23} molecules. This is a very big number. Make certain that you know how to use your calculator when using very large numbers.**

(c) 194 (**2 marks**)

🎯 **You should show all your working out for a difficult relative formula mass, just in case you make a silly mistake. The examiner can see the mistake and possibly still give you 1 mark.**

(d) $\dfrac{56}{194} \times 100$ (**1 mark**)

28.9 (**1 mark**)

🎯 **Do not use all the figures shown on your calculator: normally three significant figures will do.**

Q2 How to score full marks

(a) $\dfrac{0.69}{69} = 0.01$ (**1 mark**)

(b) $\dfrac{0.31}{31} = 0.01$ (**1 mark**)

(c) GaP (**1 mark**)

🎯 **Notice that in this calculation the numbers of moles of both elements are the same. This means that there must be the same number of atoms of gallium as there are atoms of phosphorus.**

Q3 How to score full marks

(a) $CuCO_3 \rightarrow CuO + CO_2$ (**1 mark**)

(b) (i) $CuCO_3 \rightarrow CuO + CO_2$

	1 mole	1 mole	(**1 mark**)
	124 g	80 g	(**1 mark**)
So	1.24 g	0.80 g	(**1 mark**)

🎯 **Try to do all calculations using the same method. Each step is worth a mark. The final answer should be given a unit, but there is no need for units in the working out.**

(ii) $\dfrac{0.72}{0.80} \times 100$ (**1 mark**)

90% (**1 mark**)

🎯 **Remember the equation for percentage yield as am over pm (actual mass over predicted mass), but make sure you multiply by 100 to get a percentage. You must select the correct figures to substitute into the equation. If you get a percentage yield of above 100 then you have probably got the numbers the wrong way round. You would also get a mark for quoting in words the correct equation for percentage yield.**

(iii) moles of carbon dioxide made = 0.01

so volume is $24000 \times 0.01 = 240 \, cm^3$ (**1 mark**)

🎯 **From the equation, 1.22 g is 0.01 moles of copper(II) carbonate, so 0.01 moles of carbon dioxide is made. You are not expected to remember the value of $24000 \, cm^3$, it will be included in the exam paper.**

Q1 How to score full marks

(a) As the resistance wire in the circuit gets longer its resistance increases (**1 mark**) and the current decreases (**1 mark**). This dims the bulb.

🎯 **This question asks you to explain, so you need to make two relevant scientific points to get the 2 marks.**

(b) The resistance does not increases evenly with length (**1 mark**). When the wire in the circuit is short, it will heat up (**1 mark**) causing a higher resistance than expected (**1 mark**).

🎯 **Often, with resistance of a wire or filament bulb, the resistance changes with temperature. You need to be on the look out for a chance to explain this idea because it is quite common.**

(c) (i) power = current × voltage (**1 mark**)
Recall of correct formula

power = 2 × 12 (**1 mark**)
Correct substitution

power = 24 (**1 mark**)
Correct calculation

unit is Ω or Watts (**1 mark**)
Correct unit

(ii) resistance = voltage / current (**1 mark**) **Recall of correct formula**

resistance = 12 / 2 (**1 mark**)
Correct substitution

resistance = 6 (**1 mark**)
Correct calculation

unit is Ω or Ohms (**1 mark**)
Correct unit

Q2 How to score full marks

(a) (i) The paint is charged (**1 mark**) positive. These charges repel and break up the droplets (**1 mark**), making the spray finer so less paint is used (**1 mark**). The car is given an opposite charge (**1 mark**) which is negative. This attracts the paint to inaccessible parts of the car (**1 mark**) giving a better coverage (**1 mark**).

🎯 **The question asks you to explain why it is useful _and_ how it works. You will need to cover both to have a chance of getting full marks. You must avoid writing vague things like 'the static electricity builds up' or 'the paint sticks' because paint tends to stick whether it is charged or not.**

(ii) There is friction between the fuel and the aircraft or pipe (**1 mark**). This causes charge to build up (**1 mark**) and a spark may occur causing an explosion (**1 mark**).

🎯 **You need to concentrate on explaining how the plane becomes charged for 2 of the marks. There is 1 mark for the consequence 'a spark causes an explosion'.**

(b) The paper moves towards the strip (**1 mark**) because the strip becomes charged (**1 mark**) and the paper's charges become polarised (**1 mark**).

(c) The carpet and Eli's slippers are insulators (**1 mark**). There is friction between the carpet and her slippers (**1 mark**). Charge builds up on her (**1 mark**) and flows through her into the metal radiator (**1 mark**).

🎯 **This sort of question can be answered too colloquially (unscientifically) to convince the examiner that you know any science. A candidate who wrote that 'Eli gets full of static electricity and she gets a shock off the earthed metal radiator' would be unlikely to get any marks at all. The 'static electricity' reference is too colloquial and the reference to the radiator merely restates the information in the question. Use the information in a question but do not simply restate it.**

15 Electromagnetism

Q1 How to score full marks

(a) (i) A motor has a coil (**1 mark**) that carries current (**1 mark**), making it magnetic (**1 mark**). This is affected by the field of the permanent magnet and the forces make it spin (**1 mark**).

(ii) You can use any **three** of these points. The motor will spin faster if there are stronger magnets (**1 mark**), more turns on the coil (**1 mark**), more current in the coil (**1 mark**) or a soft iron core in the coil (**1 mark**).

🎯 **These are easy ideas, often asked and well worth knowing. Some syllabuses may ask you to write about the current and force directions. You will need to learn about them if this is the case.**

(b) (i) Inside the generator there is a magnet (**1 mark**) and a coil (**1 mark**). The coil spins (**1 mark**) inside the magnetic field (**1 mark**) and electricity is generated.

(ii) You must mention **three** of these points. The generator will produce more current with stronger magnets (**1 mark**), more turns on the coil (**1 mark**), quicker movement (**1 mark**) or a bigger area in the coil (**1 mark**).

🎯 **These are also easy ideas well worth knowing. Some syllabuses may ask you to write about the current and force directions for generators. You will need to learn about them if this is the case.**

(c) (i) There are **two** ways of giving this answer. Step-up transformers increase the voltage (**1 mark**) and they have more turns on the secondary coil (**1 mark**). Put another way, step-down transformers decrease the voltage (**1 mark**) and they have fewer turns on the secondary coil (**1 mark**).

🎯 **The question asks you to make a *comparison* for 2 marks. You could write a paragraph on this but there are only 2 lines and 2 marking points available. You need to be concise and state the *two most important* differences.**

(ii) Transformers can get hot and/or make a buzzing (vibration) noise (**1 mark**). Because this noise and heat require energy (**1 mark**), less energy is kept as electricity in the secondary coil (**1 mark**).

Q2 How to score full marks

(a) the ratio of voltages = the ratio of turns (**1 mark**)

(b) either $\dfrac{V_s}{V_p} = \dfrac{N_s}{N_p}$ or $V_s = \dfrac{N_s \times V_p}{N_p}$ (**1 mark**)

$V_s = \dfrac{360 \times 230}{6900}$ (**1 mark**)

$V_s = 12\,\text{V}$ (**1 mark**)

🎯 **In this question you get 1 mark for writing down the correct formula, 1 mark for substituting the correct numbers and the third mark for the correct calculation. Some people may remember the formula the other way round:**
$\dfrac{V_p}{V_s} = \dfrac{N_p}{N_s}.$

This does not make a difference to the answer. Try it and see.

(c) Transformers need a changing magnetic field to work (**1 mark**). The a.c. current changes direction and so the field changes (**1 mark**).

(d) You can use any **two** from the following. They need to be transmitted at high voltage because that is when the current is small (**1 mark**). This means that thinner cables can be used (**1 mark**) and there is less energy transferred from the cables as heat (**1 mark**). This also reduces costs (**1 mark**).

🎯 **This question could be answered just as well by *explaining* the equation: power loss = current^2 × resistance. For example, if you halve the current the power loss is four times less and at high voltages the current is less.**

Q1 How to score full marks

(a) distance = $\frac{1}{2} \times 10 \times 5$ (**1 mark**)

    distance = 25 m (**1 mark**)

🎯 **You can work out the distance travelled by calculating the area underneath the graph. Remember that the area of a triangle is $\frac{1}{2} \times$ base $\times$ height.**

(b) average speed = distance/ time (**1 mark**)

    average speed = 25/10 (**1 mark**)

    average speed = 2.5 m/s (**1 mark**)

🎯 **In this question you need to use your answer for (a). If you got (a) wrong you can still get full marks on (b). This is called an error carried forward.**

(c) acceleration = change in speed/ time (**1 mark**)

    acceleration = 5/10 (**1 mark**)

    acceleration = 0.5 m/s^2 (**1 mark**)

🎯 **Many candidates when faced with 5/10 = 0.5 will do the opposite and divide 10 by 5 = 2. Although this is temptingly easier, it is wrong. Always take your calculator into science exams and use it correctly. Also, be sure you know the correct units, m/s^2.**

(d) force = mass × acceleration

    mass = force/acceleration (**1 mark** for rearranging the correct formula)

    mass = 3500/0.5 (**1 mark**)

    mass = 7000 kg (**1 mark**)

🎯 **As above, in this question you need to use your answer for (c). If you got (c) wrong you can still get full marks for (d).**

Q1 How to score full marks

(a) Water particles need energy to evaporate (**1 mark**). They collect this extra energy from the water in the pot (**1 mark**) and the water and bottle cool. The higher (faster moving) energy particles leave the pot leaving the low energy particles behind (**1 mark**).

🎯 **You must convince the examiner of the above points and try to avoid restating the question. There are never any marks available for this technique!**

(b) The infra-red radiation (**1 mark**) is reflected away from the aluminium foil (**1 mark**) but is absorbed by the black polythene (**1 mark**).

🎯 **Avoid saying that the radiation 'bounces' or is repelled by the shiny foil or, just as bad, that the black polythene 'attracts' the radiation.**

Q2 How to score full marks

(a) Any **three** of the following ideas. The vibrating particles in the hot-plate pass their kinetic energy onto the steel particles which vibrate more (**1 mark**). These vibrations are passed through the steel (**1 mark**). Water particles gain extra kinetic energy when they hit the steel (**1 mark**) and move around quicker (**1 mark**). These fast moving water particles hit the potato and make the potato particles vibrate more (**1 mark**). These vibrations are passed through the potato (**1 mark**) which cooks it.

🎯 **Remember that conduction takes place in solids and convection in liquids and gases. You need to explain how the particles move or try to move. They vibrate in solids but move around in liquids. Also be sure never to talk about heat particles – they don't exist!**

(b) Energy output from the pan is reduced by the lid (**1 mark**). Less energy input is needed (**1 mark**) to keep the potatoes at the cooking temperature (**1 mark**).

🎯 **When the temperature is steady, input energy = output energy. If the output energy is reduced (by using a lid) the input energy (from the hot-plate) can be reduced. This reduces both energy wastage and costs. The food does not cook quicker but costs less to cook.**

(c) (i) energy = power × time (**1 mark**)
energy = 2000 W × 1200 s (**1 mark**)
energy = 2 400 000 (**1 mark**)
units j (**1 mark**)

(c) (ii) units = power (kW) × time (hours)
units = $2 \times \frac{1}{3}$
(hint: 20 minutes = $\frac{1}{3}$ of an hour)
units = 0.67 (**1 mark**)
cost = units × 10p
cost = 0.67 × 10
cost = 6.7p (**1 mark**)

🎯 **Be sure you remember your equations and what units we use to measure things. Sometimes the question may have unhelpful units, as in this case. Time in minutes can be easily converted to $\frac{1}{3}$ of an hour, but do this before you start the calculation to avoid getting mixed up.**

18 Waves

Q1 How to score full marks

(a) (i) gamma (**1 mark**)

(ii) gamma (**1 mark**)

(iii) sound (**1 mark**)

(b) All these waves travel at the same speed (the speed of light) (**1 mark**).

> You will nearly always be tested on the electromagnetic spectrum so it is well worth learning. Lots of the questions will rely on simple recall and these are easy marks to get. Others questions, like (c) below, rely on more understanding.

(c) The microwaves are absorbed by water (**1 mark**) in the food and they cause the water molecules to vibrate more (**1 mark**).

> The frequency or wavelength of microwaves is just the right size to make water molecules vibrate. This increases the temperature of the water in the food and this energy is passed through the food by conduction in solids (meat, potatoes) but by convection in fluids (soups, coffee). Try to avoid saying that the microwaves cook the food from the inside – this is untrue, although you will hear lots of adults saying it.

(d) The gamma rays are always focused on the cancer throughout a revolution (**1 mark**) but only hit healthy cells once each revolution (**1 mark**).

> This question is a challenging one. The scientific ideas are quite difficult and even if you understand them they are difficult to put into words. In questions like this you could try to get full marks by answering in note form or even bullet points.

Q2 How to score full marks

(a) The waves are passed into the body from outside (**1 mark**) and are absorbed by soft tissue (**1 mark**) but reflected by harder tissue (**1 mark**) and detected by a sensor outside the body (**1 mark**).

(b) Ultrasound waves are 'fired' towards the kidney stones from outside the body (**1 mark**). There is no need for surgery (**1 mark**). The waves make the stones resonate (vibrate) and they break up (**1 mark**).

> You need to avoid being too colloquial with these types of questions, for example 'the waves see inside the body' or 'the ultrasound works like an X-ray' would get no marks.

(c) (i) wave speed = frequency × wavelength
$$s = f \times W \text{ (\textbf{1 mark})}$$
$$W = s/f \text{ (\textbf{1 mark})}$$
$$W = 1\,500/30\,000 \text{ (\textbf{1 mark})}$$
$$W = 0.05 \text{ (\textbf{1 mark})}$$
metres (**1 mark**)

> This is one of those equations you will often be tested on. It is well worth your while learning, understanding and practising this one.

(ii) Ultrasound has a very high frequency (**1 mark**) that is too high for the doctors to hear (**1 mark**).

Q3 How to score full marks

S waves are transverse (**1 mark**), pass through solid rock but **not** liquid rock (**1 mark**). P waves are longitudinal (**1 mark**), pass through solid rock **and** liquid rock (**1 mark**). Station A detects only p waves or does not detect s waves (**1 mark**) so the outer core must be liquid (**1 mark**).

19 The Earth and Beyond

Q1 How to score full marks

🎯 **Make sure you know the positions of the planets. If you do you can gain an easy 3 marks here.**

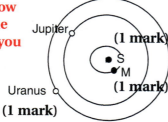

Jupiter (1 mark)

S
M

Uranus (1 mark)

(1 mark)

Q2 How to score full marks

(a) Geostationary satellites have a high orbit **(1 mark)** and stay over the same part of the Earth as it rotates **(1 mark)**. They rotate with the Earth so that TV signals can be received from and transmitted to certain parts of the Earth or other satellites **(1 mark)**.

🎯 **If you know what a geostationary orbit is you can use common sense to suggest its importance for the third mark.**

(b) Weather satellites need to move around the outside of the Earth to monitor the weather every few hours **(1 mark)**. Weather satellites need to orbit low to get good pictures **(1 mark)**.

(c) The Earth's natural satellite is the Moon **(1 mark)**.

(d) Days are caused by the rotation of a planet. The Earth rotates leaving parts of the Earth in sun or shadow **(1 mark)** once every 24 hours **(1 mark)**.

(e) The seasons summer and winter are caused by the angle of the Sun **(1 mark)** as the Earth is tilted in orbit **(1 mark)**. At low angles, the Sun's energy is diluted over the surface causing less heating effect. When it is overhead it can be very hot indeed **(1 mark)**.

🎯 **This idea is definitely worth practising as it comes up frequently on examination papers.**

Q3 How to score full marks

(a) The Big Bang theory **(1 mark)** says that the universe is expanding. There are 3 main groups of evidence for this. Light from distant stars and galaxies is red-shifted (i.e. they are moving away quickly) **(1 mark)**. Light from extremely distant stars and galaxies is red-shifted even more (i.e. they are moving away very quickly indeed) **(1 mark)**. Low frequency radiation seems to come from all parts of the universe **(1 mark)**.

🎯 **This is the most commonly held view of the Universe and therefore the one most likely to be asked about. Learn the key ideas about it.**

(b) A light year is the distance **(1 mark)** travelled by light **(1 mark)** in one year **(1 mark)**.

🎯 **Do not write 'it is the time it takes'. This will get no marks.**

(c) (i) A huge cloud of dust and gas **(1 mark)** is pulled together **(1 mark)** by gravity **(1 mark)** and its energy is transferred, through heat, raising the temperature **(1 mark)**. At very high temperatures hydrogen nuclei undergo nuclear fusion **(1 mark)** turning to helium nuclei and giving out vast amounts of energy (heat and light) **(1 mark)**. The heat causes a pressure outwards and for a long time this is balanced by the gravitational force inwards **(1 mark)**. In time, the hydrogen starts to run out and the star cools **(1 mark)** a little and grows to become a red giant **(1 mark)**. It then begins to cool further, shrinks to become a white dwarf **(1 mark)** and then becomes dimmer, when it is called a black dwarf **(1 mark)**.

🎯 **Any 6 of the marking points indicated above will score full marks. They must be written down in the correct order.**

(c) (ii) For a very large star, after the red giant phase they can start to glow brightly again as they undergo more fusion **(1 mark)**. In time, after expanding and contracting repeatedly **(1 mark)**, they explode in a supernova **(1 mark)**. These can form neutron stars and then even black holes **(1 mark)** or instead they may form a second generation star **(1 mark)** like our Sun with its own solar system.

🎯 **Most of the time you will be asked about second generation stars like our Sun. Make sure, though, that you know what happens to first generation stars. The cycle of a first generation star is the same up to the formation of a red giant. You would gain full marks for writing down any 3 of the points indicated, but they must be in the correct order.**

20 Radioactivity

Q1 How to score full marks

(a) 120 60 30 15 0 0

🎯 **In this question you get marks for interpreting information and translating it into graphical form.**

(b) All points plotted correctly (**1 mark**).

🎯 **Graphs are usually marked correct within an accuracy of +/– half a square. The examiner must be able to see the points, so do not cover them up. Draw your line in thin pencil and make sure the points are clear.**

(c) Smooth curve to the points (**1 mark**).

🎯 **Marks may not be given for a graph line that is jagged (dot to dot), a line thicker than half square, or a line that is scrappy or multiple lines.**

(d) (i) From the results beta radiation does not appear to pass through 7 mm+ aluminium so it will not penetrate 20 mm (**1 mark**).

🎯 **This question gives you the opportunity to apply your scientific understanding in making decisions. Information is given in the graph to help you.**

(ii) Gamma radiation could be used to check the thickness of a 20 mm sheet (**1 mark**).

🎯 **Remember that alpha would hardly penetrate aluminium at all.**

(iii) Half-life is the time it takes for a radioactive material to become half as (**1 mark**) radioactive (**1 mark**).

🎯 **This is much easier to write than 'half-life is the time it takes for the mass of a radioactive isotope to be reduced by a half'. By using a more complicated sentence , you run a greater risk of getting your words mixed up.**

(iv) Half-life = 5.3 years and the starting mass (at 0 years) = 20 mg. When the time = 0 years, the mass is 20 mg (0 half-lives). When the time = 5.3 years, the mass is 10 mg (1 half-live the mass is halved). When the time = 10.6 years, the mass is **5 mg** (**1 mark**) (2 half-lives the mass is halved again).

(v) The radioactivity would reduce too quickly (**1 mark**) and the machine would need more regular adjustments or (unadjusted) the sheet would appear thicker than it is (**1 mark**).

🎯 **This knowledge is not likely to have been taught by your teacher. This is because it is a problem-solving type of question. You are expected to apply your understanding of half-live to a new situation. It is a higher level question requiring imagination beyond the scope of a syllabus.**

Q2 How to score full marks

(a) beta (**1 mark**)

(b) helium nucleus (**1 mark**)
    zero (**1 mark**)

(c) Ionising means that electrons have been removed or added (**1 mark**) from an atom and the atom becomes charged (**1 mark**).